From Atoms to Patterns
Crystal Structure Designs from the 1951 Festival of Britain

The Story of the Festival Pattern Group

Lesley Jackson

Richard Dennis Publications in association with Wellcome Collection

RICHARD DENNIS
2012

To the memory of my much-loved sister,
Sue Jackson (1959-2007), Artist and Gardener

**wellcome
collection**

Published in conjunction with the exhibition:
From Atoms to Patterns
Crystal Structure Designs from the 1951 Festival of Britain
Wellcome Collection, The Wellcome Trust, London
24 April – 10 August 2008

Production by Sue Evans
Print and Reproduction by Flaydemouse, Yeovil, Somerset
Published in 2008 by Richard Dennis Publications, The New Chapel, Shepton Beauchamp, Somerset TA19 OJT
Reprinted 2012
© 2008 Richard Dennis and Lesley Jackson
ISBN: 978 0 9553741 1 1
All rights reserved
British Library Cataloguing-in-Publication Data. A catalogue record for this book is available from the British Library

Cover designed by Alan Powers
Front cover: *Afwillite 8.45 dress fabric, designed by S.M. Slade for British Celanese*
Back cover: *Beryl 8.9 plate, designed by Peter Wall for Wedgwood*
Contents page: *Haemoglobin 8.26 diagram, crystallographer Max Perutz*

CONTENTS

China Clay 8.6 tie, designed by George Reynolds for Vanners & Fennell.

CHAPTER ONE

FROM ATOMS TO PATTERNS: THE STORY OF THE FESTIVAL PATTERN GROUP

'For the first time textile designers have found their inspiration in a world that the eye cannnot see, the world of the atom and the molecule. For the first time their prompters have been the people of science.'
'The Design Story of the Century', *British Textiles*, April 1951

FIRST INITIATIVE: PATTERN IN CRYSTALLOGRAPHY

On 20 February 1946 Dr. Helen Megaw, a crystallographer from Birkbeck College, London [fig. 1], wrote to Marcus Brumwell, director of the Design Research Unit, with an intriguing proposition: 'I should like to ask designers of wallpapers and fabrics to look at the patterns made available by X-ray crystallography. I am constantly being impressed by the beauty of the designs which crop up... without any attempt of the worker to secure anything more than clarity and accuracy.' Megaw proposed that various types of crystallographic data might be used for inspiration, including X-ray diffraction photographs of crystals, as well as diagrams of atomic structures. 'I would like to suggest not merely that designers look through it for new ideas, but that they should select a few of the best which would be utilisable without substantial alteration, apply them to appropriate fabrics, and give such pattern its correct name – just as the William Morris patterns were called after their constituent flowers. I think the combination of really attractive pattern with the assurance of scientific accuracy would win a lot of attention.'[1]

Five years later, at the Festival of Britain, Megaw's ideas came to fruition - exactly as originally envisaged - in the work of the Festival Pattern Group. [figs. 2, 3, 4] Ultimately though, it was the recently established government agency, the Council of Industrial Design, which spearheaded the initiative rather than the DRU, with Megaw as scientific consultant. Brumwell had been genuinely enthusiastic: 'I have always thought that art and science can help each other and should co-operate,' he declared.[2] He even took it upon himself to sound out the sculptor Barbara Hepworth; she too was extremely positive and encouraging: 'I think it's a marvellous idea to use these "designs" for fabrics – such a good idea that I feel there ought to be money in it & you ought to lock the idea up safely in case it's pinched... / The main point seems to me to produce them as suggested in series – with their proper names – exactly as they really are. To me they are more beautiful than any man-made pattern.'[3]

Alastair Morton, the enlightened art director of avant-garde textile firm Edinburgh Weavers, was suggested as a potential manufacturing partner. He too was reported to be 'extremely interested' and attempts were made to set up a meeting, although it appears this never happened. Meanwhile, Megaw was invited to write a monograph on the subject to be published by the DRU. She duly submitted an essay - 'Pattern in Crystallography' – on 13 November 1946. However, from this point onwards the DRU began to prevaricate, informing her after several months' delay that publication had been deferred. Eventually, three years later, they conceded that the book had been dropped

fig.1 Dr Helen Megaw, 1951.

fig.2 Guide to Exhibition of Science, Science Museum, 1951.

fig.3 *Exhibition of Science: The Structure and Mechanism of Life. Afwillite 8.45 wallpaper by John Line on screens; light fitting designed by Brian Peake, produced by GEC.*

fig.4 *Festival of Britain: Dome of Discovery and Skylon.*

SCIENTIFIC CONTEXT: X-RAY CRYSTALLOGRAPHY

By this date Megaw had moved back to Cambridge to work at the Cavendish Laboratory, where she had originally done her PhD in the early 1930s with J.D. Bernal, latterly collaborating with him again at Birkbeck where he was now Professor of Physics. Bernal was a towering figure in the scientific world and played a pivotal role among crystallographers as a mentor and guru. Sir Lawrence Bragg, who now ran the Cavendish, was another legendary scientist who, with his father, Sir William Henry Bragg, had invented the new science of X-ray crystallography, for which they were jointly awarded the Nobel Prize for Physics in 1915. Physics was central to crystallography, but the subject also encompassed chemistry, biology, mineralogy and mathematics. For scientists, crystallography had great appeal, enabling them to study the structure of matter at a sub-microscopic level and - for the first time - to work out the arrangement of atoms within molecules.

In her essay, 'Pattern in Crystallography', Megaw explained the allure of X-ray crystallography: 'Most people have at one time or another noticed and admired the "frost flowers" traced on window-pane or pavement by the freezing of thin films of moisture or water-vapour. It is easy to understand that these patterns owe their beauty to the underlying regularity of the crystal structure...This inner regularity is not confined to materials which can grow into recognisable crystals. It is possessed by almost all solids, including substances like chalk and jewellers' rouge, which look to the eye like amorphous powders. It is this pattern, this regularity in the arrangement of atoms, which the X-ray crystallographer sets out to investigate.'[4]

The discovery that all matter is composed of a mass of tiny interlocking crystals dates back to the 18th century. However, it was not until X-ray diffraction was applied to crystallography by the Braggs from 1912 onwards that crystal structures could be effectively explored. The process involves taking X-ray photographs of crystals or crystalline materials, either naturally occurring or artificially created [fig. 5], as Megaw outlines in her essay: 'A beam of X-rays falls on the crystal, which, because of the regularity of its repeated pattern, diffracts it in a regular way, and the diffracted rays are recorded as black spots or lines on a photographic film... From the position and intensity of the spots and lines on the photographs it is possible to calculate what the size and contents of the unit cell must be.'[5] [fig. 6] These calculations were used to plot 'maps' indicating the relationships between atoms. [figs, 7,8]

It was Megaw's instinctive appreciation of the aesthetic qualities of these diagrams which led her to propose their application to pattern design. Interestingly though, right from the early days the Braggs themselves had used the analogy

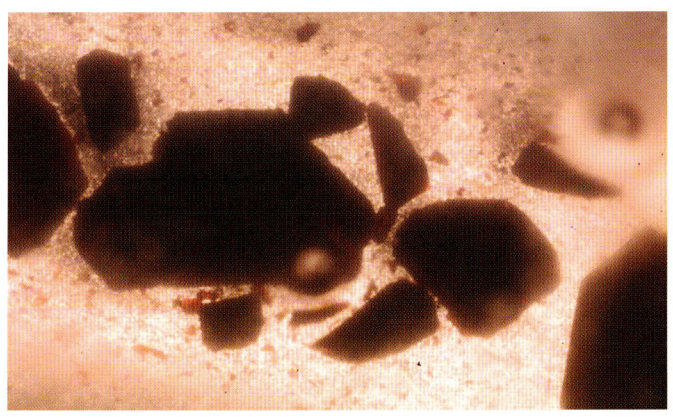

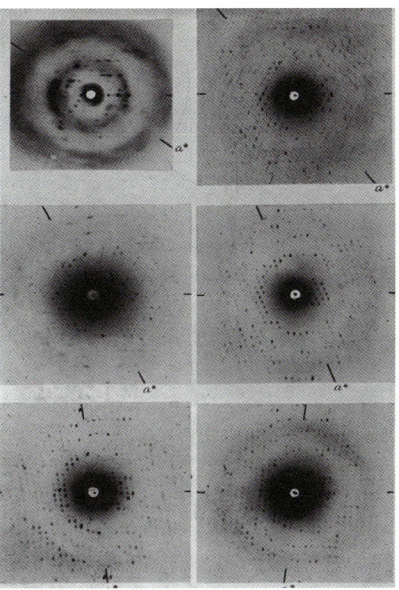

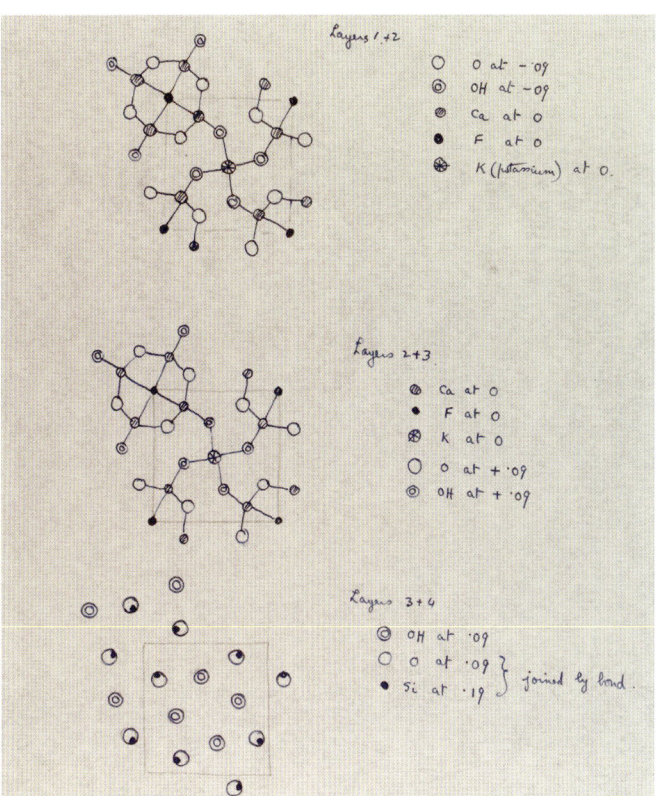

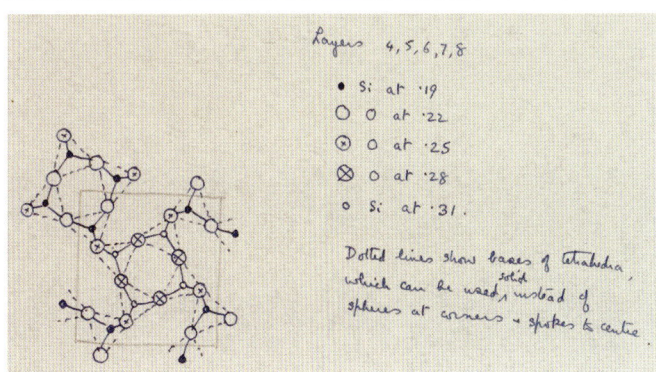

Top left, fig.5 *Haemoglobin crystals.*

Bottom left, fig.6 *5° oscillation photographs of horse methaemoglobin crystals in six states of swelling. From J. Boyes-Watson, E. Davidson, M.F. Perutz, 'An X-ray study of horse methaemoglobin. I', Proceedings of the Royal Society, Series A, 26 September 1947, vol.191, no.1024, p.128, pl.6.*

Above, figs.7-8 *Working drawings by Helen Megaw showing layers of atoms in crystal structure of apophyllite.*

of wallpapers to explain the principles of crystallography, a metaphor subsequently adopted by their followers: 'A crystal structure, like a wallpaper, consists of a unit of pattern which repeats itself indefinitely,' writes Megaw [fig. 9], arguing that it is not only apt, but logical, that crystallography should, in turn, inspire wallpaper. 'If our path to the understanding of crystal structures has been made easier for many of us at its outset by the contemplation of wallpapers, we now, having explored further into the territory opened up, can bring back new material which we hope may be built up into fresh advances in fabric design.'[6]

fig.9 *Insulin 8.27 dyeline print, diagram by Dorothy Hodgkin.*

At the Cavendish Laboratory, Lawrence Bragg built up a world-class crystallographic research team, encompassing not only minerals specialists, such as Megaw, W.H. Taylor and Bragg himself, but biophysicists such as Max Perutz and John Kendrew, who were just venturing into the emergent field of molecular biology. They were joined a few years later by Francis Crick and James Watson who, with the help of X-ray diffraction photographs taken by Rosalind Franklin at King's College, London, worked out the double helix structure of DNA in 1953. Having weathered the difficult wartime years, there was a great feeling of excitement and optimism amongst scientists during the late 1940s. X-ray crystallography was one of the most significant and stimulating branches of science at the time – the new frontier - and Britain was at the cutting edge of international research. As Assistant Director of Research at the Cavendish, Megaw was at the epicentre of the crystallographic scene. She had close links with all the key protagonists, including Dorothy Hodgkin from the Department of Chemical Crystallography at Oxford and Kathleen Lonsdale, Reader in Crystallography (and soon to be professor) at University College, London.

MEETING OF MINDS: HELEN MEGAW AND MARK HARTLAND THOMAS

In November 1948 Megaw received a letter from Kathleen Lonsdale asking to borrow some of her crystal structure diagrams: 'I have rashly promised to give a (semi-popular) lunch-hour lecture here ... on "Art and Architecture in Science,"' she wrote. 'I wondered if perhaps you could lend me some of the beautiful patterns you have, so that I could have slides made of them?'[7] Megaw obliged, and Lonsdale later re-used the slides in a lecture about crystallography to the Society of Industrial Artists during a weekend course at Ashridge the following May. On this occasion Mark Hartland Thomas (1906-1973), Chief Industrial Officer from the Council of Industrial Design, was in the audience. An architect by profession who had joined the COID in 1947, Hartland Thomas immediately realised the potential of this material. Writing to Megaw on 1 June 1949, he laid his cards on the table: 'The idea of the Conference was to show to industrial designers, aspects and patterns of things that are of interest to scientists and mathematicians as well as artists, in order to accord to them a more modern inspiration for their work... Mrs Lonsdale showed at the end of her lecture, some transcriptions of crystallographic diagrams that she told us you had prepared with the idea that they might be used directly as decorative patterns of more than decorative interest in textile printing or pottery transfers, or the like. Some of these seem to us to be very promising, and I wonder whether you would be interested for me to see whether we can place any of these patterns with manufacturers?'[8]

Megaw was more than happy to cooperate and promptly sent Hartland Thomas some crystal structure diagrams, along with a copy of her essay: 'My original idea had been in connection with textiles;' she explained. 'I thought that if the textile kept pretty faithfully to the scientific diagram and judicious use of this fact were made in describing or advertising a series of such materials it would add to their attractiveness.'[9] It was at this point that the idea of a special design project linked to the forthcoming Festival of Britain was first floated by Hartland Thomas. [fig. 10] The Festival was intended as a platform for British ingenuity and creativity in science, technology and the arts, so a project combining all three seemed ideal. Hartland Thomas was already deeply involved with the Festival. As well as coordinating the COID's Stock List (a comprehensive register of 'approved' products used to promote 'good design' at the Festival, later perpetuated as the Design Index), he was a member of the Festival of Britain Presentation Panel responsible for overseeing buildings and displays on the South Bank. In spite of his prodigious workload, however, Hartland Thomas jumped at the opportunity of tackling this new initiative: 'My idea is to

fig.10 *Festival of Britain, South Bank, London, 1951.*

make a small group consisting of a manufacturer of each of the following, namely: dress textiles, furnishing textiles, carpets, pottery, linoleum and wallpaper, with a view to showing goods in these categories decorated with crystallographic patterns.'[10]

Although Hartland Thomas used the word 'small', it is evident right from the start that he had great ambitions for the project. Crystallography was an exciting new source of imagery, something genuinely fresh and original, he believed: 'I had it in mind that we are at a stage in the history of industrial design when both the public and leading designers have a feeling for more richness in style and decoration, but are somewhat at a loss for inspiration. Traditional patterns that have come down to us from ancient Greece and elsewhere, had lost much of their sparkle by now; and the fashionable alternative of a doodle on a piece of paper, folded for symmetry, could hardly lay the foundations of a new school of design. / But these crystal structure diagrams had the discipline of exact repetitive symmetry; they were above all very pretty and were full of rich variety, yet with a remarkable family likeness; they were essentially modern because the technique that constructed them was quite recent, and yet, like all successful decoration of the past, they derived from nature – although it was nature at a submicroscopic scale not previously revealed.'[11]

Over the next two years Hartland Thomas personally spearheaded and piloted the Festival Pattern Group - it was very much his 'baby'. Although from different professional spheres, he and Megaw made an effective team; he respected her expertise as a scientist and ensured that her high-minded aspirations for the project remained sacrosanct; she trusted in his ability to find sympathetic manufacturers and, literally, to deliver the goods. From July 1949 onwards, Hartland Thomas embarked on discussions with potential manufacturers. Although some fell into place quite quickly, others took longer to identify; new partners were still being recruited as late as February 1951. In the end, twenty-eight firms participated in the Festival Pattern Group, producing a total of eighty designs. From a purely organisational point of view, simply in terms of the volume of correspondence, this was a remarkable undertaking. But bearing in mind the severe practical constraints on manufacturers during the early post-war era – government restrictions on the sale of products on the home market, crippling purchase tax, chronic labour shortages, lack of essential materials, not to mention rationing and Utility regulations – the fact that the Festival Pattern Group happened at all at this date seems extraordinary, never mind the originality and variety of what was produced.

FORMATION OF THE FESTIVAL PATTERN GROUP

Starting with existing contacts such as Sir Thomas Barlow (former chairman of the COID and head of the textile firm, Barlow & Jones), Hartland Thomas solicited suggestions from experts in the field, such as Dennis Lennon at the Rayon Design Centre who recommended the synthetic fibres company British Celanese. Other firms on his original target list included well-known companies such as wallpaper manufacturers John Line & Sons, glassmakers Chance Brothers, rubber producers Dunlop [fig. 11] and carpet manufacturers James Templeton. While concentrating on established firms, however, he was also on the look-out for individual flair, exemplified by maverick designer Arnold Lever (formerly of Jacqmar), who had set up his own

fig.11 Advert for Dunlop from Guide to Exhibition of Science.

dress fabrics company in 1947. Each firm was carefully chosen for their prowess in a specialist field, partly out of desire to achieve maximum diversity, but also to avoid potential conflicts between commercial competitors.

Within a couple of months Hartland Thomas announced that he had secured one of the main restaurants at the Festival for 'an experiment in pattern design', with crystal structure furnishings as the theme of the décor. At one time the name 'Crystal Café' was considered, but 'Regatta Restaurant' was chosen in the end. [figs. 12, 13] By a quirk of fate the architects for the building were the Design Research Unit, led by Alexander Gibson and Misha Black. Ian Cox, Director of Science and Technology at the Festival, was thrilled at the prospect of a project combining science and design. Scientific displays were planned for two key sites: the Dome of Discovery on the South Bank and the Exhibition of Science at the Science Museum at South Kensington. [fig. 14] 'The Structure of Matter' had already been identified as one of the major themes, with X-ray crystallography as a vital part of this story, so this new initiative slotted in perfectly.

In August 1949 Megaw was officially invited to act as scientific consultant for the project - the name 'Festival Pattern Group' was not adopted until November 1949 - her role being to select and collate all the crystal structure diagrams to be used in the scheme. A lengthy debate ensued about the issue of copyright. Megaw declared: 'I maintain that these crystal structures are facts existing in the world of nature; in exactly the same way as objects which need microscopic technique to gain knowledge of them. Perhaps they could have been patented; in fact they

Far left, fig.12
Regatta Restaurant, designed by Misha Black and Alexander Gibson, Design Research Unit.

Left, fig.13
Interior of Regatta Restaurant.

have not been, but have been published openly for everyone to use as they wish.'[12] Eventually it was agreed that, in the case of diagrams formulated by one particular crystallographer, Megaw would liaise directly with the individual concerned and obtain permission for their diagrams to be used. This perhaps explains why, in addition to her own contributions, so much of the material was supplied by her colleagues at the Cavendish (Lawrence Bragg, W.H. Taylor, Max Perutz, John Kendrew and

a research student called June Broomhead), partly because they were doing such interesting work, but also because they were on the spot. Other diagrams were sourced from crystallographers Megaw already knew well, or well enough to approach, mostly at other universities (Dorothy Hodgkin at Oxford, Professor Gordon Cox and Dr. G.W. Brindley at Leeds, Professor John Monteath Robertson at Glasgow), but also in industry (Charles William Bunn and Myra Bailey at ICI).

fig.14 Exhibition of Science, Science Museum: entrance screen designed by Gordon Andrews on theme of the peaceful application of atomic energy.

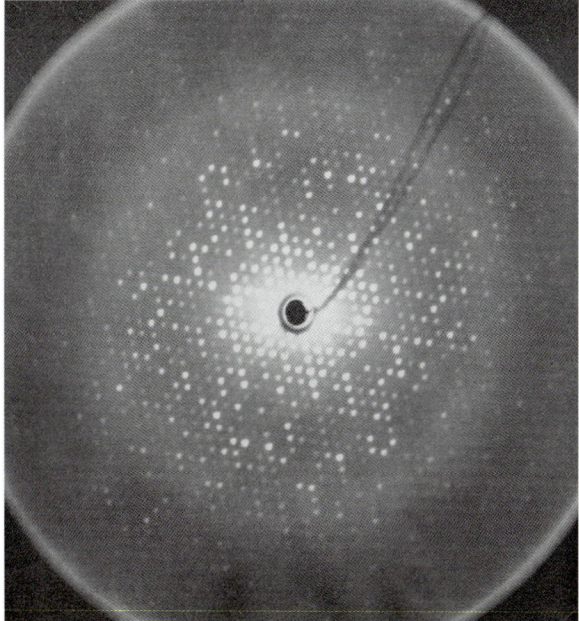

Right, fig.15 *X-ray diffraction pattern from a haemoglobin crystal.*

Far right, fig.16 *Diagram from M.F. Perutz, 'An X-ray study of horse methaemoglobin. II', Proceedings of the Royal Society, Series A, 3 February 1949, vol.195, no.1043, p.478, fig.1.*

PIONEER CRYSTALLOGRAPHERS: MAX PERUTZ, JOHN KENDREW AND DOROTHY HODGKIN

With hindsight Megaw's choice of crystallographers reveals amazing astuteness and prescience, including three scientists who would go on to win the Nobel Prize: Max Perutz and John Kendrew in 1962, Dorothy Hodgkin in 1964. All three were working in the field of proteins, the newest and most difficult and elusive area of crystallographic research. In 1947 Perutz had been appointed head of a new experimental section at the Cavendish funded by the Medical Research Council called the Unit for the Study of the Molecular Structure of Biological Systems (later renamed Molecular Biology Research Unit). Perutz went on to run the world-famous Laboratory of Molecular Biology (LMB), which split off from the Cavendish in 1962, but back in 1949 his area of research was still so embryonic that it did not even have an established name.

Perutz's interest lay in haemoglobin, the vital oxygen-carrying protein molecule in the blood, which he studied in crystalline form, mainly using horse blood. [figs. 15, 16] He

was also working on horse methaemoglobin, a non-oxygen-carrying variant of haemoglobin. In 1949 Perutz gave copies of two recent crystal structure diagrams to Megaw, identified in the Festival Pattern Group list as Horse Methaemoglobin 8.23 and Haemoglobin 8.26. [fig. 17] Haemoglobin 8.26 was one of the first diagrams to be circulated to manufacturers; Sir Thomas Barlow referred to it as his favourite in November 1949, and his firm eventually produced four fabrics based on this design (FPG 1-3). Haemoglobin 8.26 has added significance as it became by far the most widely interpreted pattern in the scheme, applied to a diverse range of printed and woven dress and furnishing fabrics, as well as lace, tie silks, leathercloth, plastic laminate, wallpaper, wrapping paper and ceramics. [fig. 18] From a scientific point of view this diagram is intriguing as it was never published, whereas almost all the others were.[13] It would take another ten long years of painstaking work before Perutz finally succeeded in determining the full structure of haemoglobin in 1959, so there is a special resonance to this simple early diagram.[14]

Far left, fig.17 *Haemoglobin 82.6 dyeline print, diagram by Max Perutz.*

Left, fig.18 *Haemoglobin 8.26 dress fabric by Barlow & Jones; and Haemoglobin 8.26 plate by Peter Wall.*

fig.19 *John Kendrew and Max Perutz with second myoglobin structure model ('forest of rods'), 1959.*

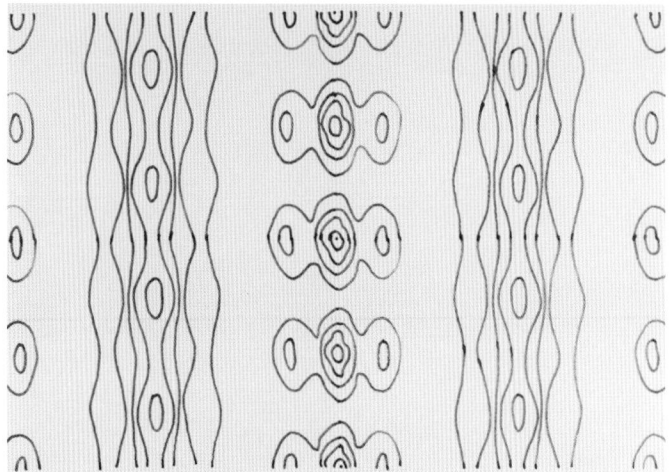

fig.20b *Myoglobin 8.46g dyeline print, diagram by John Kendrew. For scientific source, see fig 20a, bottom left.*

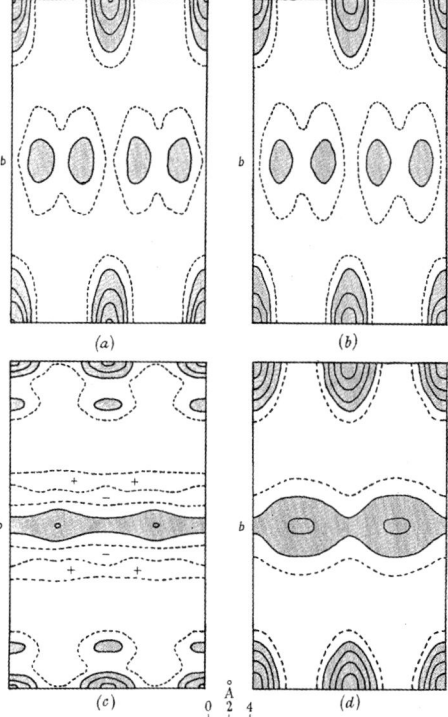

FIGURE 22. Model [20Ī] Patterson projection of myoglobin using (a) $2_{14} \cdot \frac{1}{3}$ chains, (b) $2_{13} \cdot \frac{1}{3}$ chains, (c) pairs of $2_7 b$ chains, (d) pairs of 3_{10} chains.

fig.20a *Scientific source for John Kendrew's Horse Myoglobin 8.46f-g diagrams: W.L. Bragg, J.C. Kendrew, M.F. Perutz, 'Polypeptide Chain Configurations in Crystalline Proteins', Proceedings of the Royal Society, Series A, 10 October 1950, vol.203, no.1074, p.353, figs. 22a-d.*

The first person to solve a protein crystal structure in the end was Perutz's colleague, John Kendrew, who worked alongside him in the molecular structures unit and succeeded in determining the structure of myoglobin in 1957. [fig. 19] Myoglobin is an oxgygen-carrying protein molecule active in muscle tissue. Kendrew studied both horse myoglobin and whale myoglobin, although horse myoglobin ultimately proved unproductive in scientific terms. Diagrams of both were passed to Megaw some time before January 1950, several of which were subsequently published. [figs. 20a and b] The two whale myoglobin diagrams (8.46a-b) were not picked up by any FPG manufacturers, but three horse myoglobin diagrams (8.46c, f, g) were developed by ICI's Leathercloth Division as nitrocellulose-coated upholstery fabrics and wallcoverings. [figs. 21, 22] As with Perutz's diagrams, they are significant because, at the time, they were 'hot off the press' cutting-edge research. It was only because Megaw worked in close proximity to these two men that she was able to gain access to such valuable scientific material.

Megaw also had a special relationship with Dorothy Hodgkin. The two women had become good friends while working together in Bernal's laboratory in Cambridge between 1932-34. In fact, when Dorothy Crowfoot (as she was then) got married in 1937, Megaw presented her with a linen cushion featuring the crystal structure of aluminium hydroxide embroidered in silver, red and blue thread – her first foray into crystallographic patterns. (Later she made Christmas cards decorated with the same motif).[15] Dorothy Hodgkin moved to Oxford in 1934 and it was here that she took her first X-ray diffraction photograph of insulin crystals and pursued her ground-breaking early research on insulin. During the war she was diverted onto penicillin and she then worked on vitamin B_{12}. It was for her achievement in determining these two crystal structures that she was awarded the Nobel Prize. Hodgkin later returned to insulin and finally succeeded in unravelling its complex structure in 1969.

Her three FPG diagrams, Insulin 8.24, 8.25 and 8.27, all date from the pre-war period. [fig. 23] Insulin 8.25 was published in 1938 [fig. 24]; the other two probably date from the same period as they all closely related. Again, the reason why Megaw obtained access to this material was because of close professional and personal ties, as revealed by their correspondence: 'The

fig.21 *Telephone recess, Exhibition of Science: Myoglobin 8.46f mural rexine by ICI Leathercloth.*

fig.22 *Cinema foyer, Exhibition of Science: Myoglobin 8.46g vynide upholstery by ICI Leathercloth and Haemoglobin 8.26 curtains by Warner.*

fig.23 Insulin 8.27 Patterson contour map of interatomic distances, diagram by Dorothy Hodgkin.

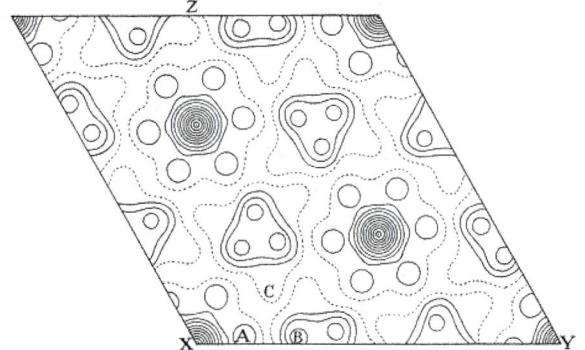

FIG. 1. P(xy) for insulin derived from Patterson-Fourier analysis. Contours at 5 units apart.

fig.26 Insulin 82.7 plastic laminate by Warerite (variant of FPG 70).

particular patterns of yours which I want to ask for are the two of which I send you prints, and a third which is very like a refined version of the second...There may well be others, since the choice is so rich that I've only picked out the nearest things to hand, and haven't started a systematic search. If you yourself have any favourites, would you either give me the references if they're published or lend me diagrams or lantern slides to copy from? At the meeting I went to, the manufacturers – or their designers – were very much taken with insulin – except for its name, which, to my surprise, they felt put people off.'[16] As Megaw predicted, Hodgkin's insulin diagrams proved very popular with FPG manufacturers, and were eventually translated into wallpapers, carpets, plastic laminates, linoleum, leathercloth, lace, ties and 'Oxvar' decorative finish. [figs. 25, 26]

Middle, fig.24 Scientific source for Dorothy Hodgkin's Insulin 8.25 diagram: Dorothy Crowfoot, 'The crystal structure of insulin. I. The investigation of air-dried insulin crystals', Proceedings of the Royal Society, Series A, 18 February 1938, Vol.164, no.919, p.594, fig. 1.

Left, fig.25 Regatta Restaurant: Insulin 8.27 wallpaper by John Line; Pentaerythitol 8.18 ashtrays by Wood Brothers on tables.

Helen Megaw's role as Adviser on Crystal Structure Diagrams

Although Megaw's connections proved advantageous in securing interesting material, clearly she also had a good eye as the diagrams she selected proved inspirational for manufacturers; there were never any complaints on that score. Indeed, had Megaw not been endowed with a keen visual sense she would never have conceived the idea in the first place. 'I always had pleasure in patterns,' she later recalled. 'I think that was one reason why I enjoyed Crystallography so much.'[17] In her essay, 'Pattern in Crystallography', she revealed her underlying motivations: 'It is often put forward as a professed aim of science to gain control of the processes of nature by learning to understand their mechanisms; but to most scientists, perhaps, an appreciation, however inarticulate, of the pattern underlying these processes is the driving force of their work. For the crystallographer these patterns are readily translatable into visual terms. It is hoped that these few examples drawn from such a rich field may suggest to designers ways in which to broadcast to a wider public some of the aesthetic pleasure found in the subject by crystallographers themselves.'[18]

On 6 January 1950 Megaw was formally appointed 'Adviser on Crystal Structure Diagrams for the Festival of Britain Exhibition' for a fee of £100 per annum, backdated to October 1949. In her contract, and indeed throughout the project, great emphasis was laid on the importance of scientific authenticity and accuracy: 'The duties will include provision of an adequate supply of crystal structure diagrams and advice to manufacturers and designers about the limits within which designs may be adapted for commercial use while retaining their scientific meaning and accuracy, and explaining the subject in general terms to those who will be engaged in designing, manufacturing, and selling the materials.'[19]

Following her appointment Megaw wrote to various crystallographers seeking permission to use their diagrams. In her letter to Professor John Monteath Robertson at the University of Glasgow, she fills him in on the background: 'Some years ago I began to get interested – as I think a good many crystallographers have – in the possibility of using crystal structure patterns for designs for textiles, wallpapers, etc. I collected a number of published patterns to illustrate the idea. Recently the Council of Industrial Design has taken it up, and is interesting manufacturers in it in connection with the Festival of Britain. I am acting as adviser and we now want to seek your permission to use some of your patterns. Whether you are willing to do so or not, may I ask you to keep the whole thing secret? They attach great importance to bringing it all out with splash on the appointed day, May 1951, and the effect would be spoiled if people got to know about it beforehand. / My part is to supply them with diagrams to choose from. They rather liked the Pattersons and electron-density maps, and indeed I myself have been quite surprised how effective a lot of these look when one draws them out to show a large number of repeats, even when the asymmetric unit looks quite undistinguished. Your work offers a very large number of possibilities to choose from, and I should like to be free to do so. / We have no guarantee that the manufacturers will stick to scientific accuracy, but they seem to recognise that it is part of the effectiveness of the idea that they should do so.'[20]

Each crystallographer was offered a fee of £5 per diagram in return for a non-exclusive licence, which most duly signed and returned. The exception was Dorothy Hodgkin, who replied saying she was happy for her diagrams to be used, but objected to signing the form on principle: 'I feel rather doubtful whether I own any copyright of a pattern perpetrated by nature.'[21] In March 1951 participants were asked to sign a second authorisation form in return for a one-off payment of £10. Again, they all complied apart from Hodgkin, prompting Megaw to explain to Hartland Thomas: 'I think it is quite possible Mrs Hodgkin will not send back this form; she didn't sign the other, and merely gave permission for her diagrams to be used, in a personal letter to me; I think she is inclined to feel she has no right to claim more money on the scientific facts, even though they were her discovery.'[22]

Scientific Sources for Crystal Structure Diagrams

All the crystallographers were sworn to secrecy, and in fact, apart from Megaw, whose role was openly acknowledged in subsequent publicity, their names were deliberately withheld at the time. The reason for this appears to have been concern (presumably on Megaw's part) to protect their scientific reputations, in particular the perceived need to separate their serious academic research from the appealing, but obviously more light-hearted, interpretations of their diagrams by the FPG. One contributor, the physicist G.W. Brindley, specifically flagged this up: 'I think we ought to be properly safeguarded against any suggestion that the diagrams are scientifically correct. It might [need] to be stated rather explicitly in any handbook which refers to the origins of the designs that the originals have been treated in a very free & easy way. I feel sure you will sympathise with this & ensure that scientific results & reputations are not sacrificed.'[23] While there is no evidence of any reluctance to cooperate on the part of any crystallographers – indeed quite the opposite; some (including Brindley) spontaneously offered additional diagrams – equally, none expressed any desire to be openly associated with the FPG or to have their names attached to particular products.

Within the crystallographic world, however, everyone would have known (or could easily have found out) who was responsible for the original diagrams, as most of them had been published in scientific journals, such as *Proceedings of the Royal Society* or *Acta Crystallographica*. Lawrence Bragg's seminal book *Atomic Structure of Minerals* (1937) was another primary source for FPG crystal structures. In the case of unpublished diagrams, their subject matter automatically linked them to known research topics by particular individuals: Hodgkin with insulin, for example, and Perutz with haemoglobin. Describing the process of selecting the diagrams, Megaw later recalled: 'I went through numbers of crystallographic journals, and unpublished papers that had come to me, picking out designs that I thought would be attractive. I traced these, and for each, joined it on to several repetitions on the same page, to emphasize the essentially repetitive character of the structure. Sometimes I added a bit of colour. These papers were then submitted to the industrial designers.'[24]

The key point about the Festival Pattern Group is the integrity and authenticity of the scientific source material, and the accuracy with which it was interpreted; these were the defining

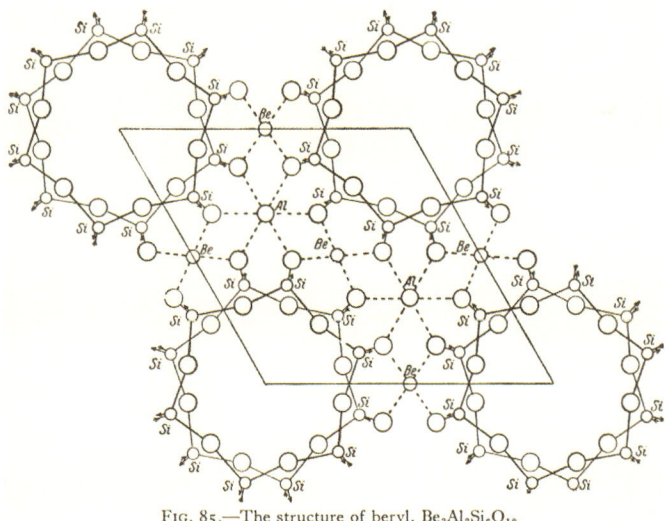

FIG. 85.—The structure of beryl, $Be_3Al_2Si_6O_{18}$

fig.27 Scientific source for Lawrence Bragg's Beryl diagram: W.L. Bragg, The Crystalline State, 1933.

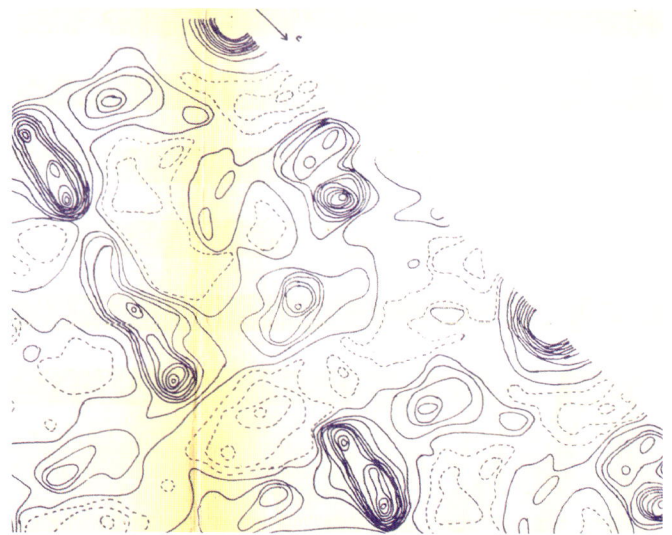

fig.28 Afwillite 8.44 dyeline print, diagram by Helen Megaw. This diagram provided the source for a woven furnishing fabric designed by Marianne Straub for Warner & Sons, used in the Regatta Restaurant.

features of the Festival Pattern Group and its fundamental raison d'être. Nothing like this had ever been attempted before in the design world, and nothing quite like it has been attempted since.[25] Chapter 5: An A-Z of Crystal Structures traces each diagram back to its original source, revealing that in most cases they relate very closely to – or are exact reproductions of - published diagrams [fig. 27], and highlighting the diverse branches of science underpinned by crystallography. Minerals made up the majority of crystal structures selected by Megaw, hardly surprising given that this was her own field, and the most widely explored field of crystallography at this date. Biological molecules formed another distinct subgroup, as did synthetic polymers, reflecting new research in these fields. In addition, various chemical compounds were included in the scheme, relating to areas such as pharmaceuticals and pesticides. Although their names sound esoteric, therefore, many FPG substances were commonplace or had everyday applications - this was very much the stuff of life.

FESTIVAL PATTERN GROUP: MODUS OPERANDI

During autumn 1949 Hartland Thomas contacted many more manufacturers and began to circulate sample diagrams to those who were interested, on the strict proviso that total secrecy was enforced. The minutes of the first FPG meeting, held on 16 December 1949, neatly summarise its aims: 'All members of the Group were acquainted with the proposal to develop for 1951 a new system of design for surface pattern, based on crystal structure patterns. There were two main projects in view; firstly the furnishing of a particular restaurant in the South Bank Exhibition with materials, tableware etc., incorporating these patterns, and secondly to have similar patterns ready for sale as widely as possible throughout the country and the world on, but not before, 1 May 1951. The scheme provided a great opportunity to promote good design and reach overseas markets at a time when the attention of the world was directed to the 1951 Festival of Britain, and was peculiarly appropriate to the spirit of the Festival, in that Britain led the world in the study of crystallography.'[26] Later a constitution was drawn up and FPG

members were asked to pay an annual subscription of £10. The COID even consulted a patent agent about registering the words 'Festival Pattern' as a trademark, but decided against this in the end.

The inaugural meeting provided an opportunity for Helen Megaw to educate manufacturers about X-ray crystallography. She stressed that 'in spite of their decorative character the patterns had all arisen in the course of scientific work and represented scientific realities.' Her slides included X-ray photographs of crystals, as well as 'patterns based on the mapping of different aspects of crystal structure; for example, the main lines of force between atoms, or relative heights, densities or distances.' It was these diagrammatic representations of crystal structure that were proposed as the basis for designs. Hartland Thomas told the assembled industrialists that, whilst recognising the diagrams would need to be adapted, he was anxious that the resulting patterns 'should not offend the scientist.' Ernest Goodale (from Warner & Sons) injected a note of caution, saying 'it would be a mistake to attach too much importance to scientific accuracy,' and emphasising that 'the first essential was that each pattern should appeal *in itself*.'[27]

Following the meeting, FPG members were sent a set of authorised diagrams, supplied as dyeline prints, plus Helen Megaw's 'Notes on Crystal Structure Diagrams' (see Chapter 4). These notes provide an invaluable explanation of the diagrams, clearly distinguishing between the two main types: ball-and-spoke structures representing atoms (depicted as circles) and the bonds between them (drawn as lines); and curvilinear contour maps recording either the distribution of matter (electron-density maps) or the distance between atoms (Patterson projections). [fig. 28] Megaw repeatedly cited the analogy of maps when explaining the diagrams to a lay audience: 'One might sum it up by saying that all diagrams of a given crystal structure, however different from one another they may appear at first sight, have the same kind of family resemblance as a series of maps of the British Isles designed to show separately such things as roads, railways, physical features, rainfall, and density of population.'[28] The map metaphor was further extended in her advice to manufacturers: 'It is legitimate to show only those features of the map which one desires to emphasise for a given purpose; but it is not legitimate to change their positions, or to put in things which are not there, or to put in some things of one kind and leave out others exactly similar. Colouring may be done in any way, provided things which are identical are coloured identically.'[29]

FESTIVAL PATTERN GROUP MANUFACTURERS AND DESIGNERS

Manufacturers were free to choose which diagrams suited them. One of the most positive aspects of the project was that, rather than importing 'star' designers, most of the design work was undertaken by in-house staff in manufacturers' design studios, bringing welcome personal recognition to individuals who would normally remain anonymous. On one occasion, however, a high-profile consultant designer was parachuted in: J. Beresford Evans at Chance Brothers, enlisted at the suggestion of Hartland Thomas after the company's own creative endeavours failed to come up to scratch. Three young ceramic designers - Peter Wall, Peter Cave and Hazel Thumpston - all became involved initially as part of a student project at the Royal College of Art, coordinated by Professor R.W. Baker. All three were subsequently recruited by the firms with whom they were paired: Wall joined the design studio at Wedgwood; while Thumpston and Cave were taken on as trainee designers at E. Brain & Co. Another ex-RCA student, Charles C. Garnier, also benefited from the scheme: he went on to become Division Designer at ICI Leathercloth. Most FPG designers were identified by name in the resulting publication, *The Souvenir Book of Crystal Designs*, although this includes several errors.[30] The COID would have preferred all designers to be credited, but some companies, such as Barlow & Jones and Old Bleach Linen Company, steadfastly refused to supply this information.

Hartland Thomas kept in close contact with each company by letter, normally dealing with one of the directors, an indication of how prestigious the project was perceived to be. FPG meetings were held every couple of months – seven in all during the first year, plus one to mark the launch of the Festival and one just after it closed. Most companies sent one or two representatives to meetings (normally either the designer, works manager, or sales director) and took along samples for discussion. An ad hoc Subcommittee on Textiles was set up to coordinate this area, as textile manufacturers constituted a third of FPG members.

Attitudes towards the project varied considerably between different companies, depending on the outlook of the person in charge. Some were cautious, such as the Huddersfield woollen firm Dobroyd, manufacturers of 'exclusive ladies' cloths'. Some were initially sceptical, including Sir Thomas Barlow, who told Hartland Thomas privately that 'he thought the whole project "humbug" but would join in for my sake.'[31] On the whole though, there was a high level of commitment amongst participants (membership was entirely voluntary after all). Some were incredibly enthusiastic, such as James Templeton, Spicers and A.C. Gill. Company designers were particularly appreciative of the opportunity to work on such an unusual project. At the third FPG meeting, it was reported that Henry Dowling, Chief Decorative Adviser at John Line, 'showed a large number of

fig.29 Advert for Warerite, Design magazine, May-June 1951.

exciting wallpaper designs, [and] expressed the satisfaction of his firm's design department at having this fresh and unusual source of inspiration.'[32]

Helen Megaw vetted the patterns from a scientific point of view, raising concerns if she thought that the diagrams were being misinterpreted. Retrospectively, some of her judgements seem rather harsh, although at the time she appears to have been much more positive. 'When I saw what they had done, my feelings were mixed,' she later recollected. 'Some were legitimate and beautiful; some were quite illegitimate and horrible; others in between.'[33] Her main concern at the time was to discourage manufacturers from taking liberties. After the fourth FPG meeting on 13 July 1950, when a large display of samples was presented, she remarked: 'I think it is important that the contour patterns should not be superposed (sic) on a background of irregular splotches of different colours, as was done with one or two of the patterns shown, because this detail would be scientifically meaningless.'[34]

As a non-scientist, Hartland Thomas was less critical. He declared at this meeting that he was extremely pleased what had been put forward so far, and heartened by the spirit of cooperation within the group. He concluded that this 'new source of pattern design had had a catalytic effect, giving new impetus to design... and that the patterns showing a free and imaginative treatment of the crystalline structure were more successful than those adhering too closely to their source.'[35] In a progress report to the COID three months later he observed: 'Development work is proceeding well. The amount of personal effort that members are putting into the project is most gratifying.'[36]

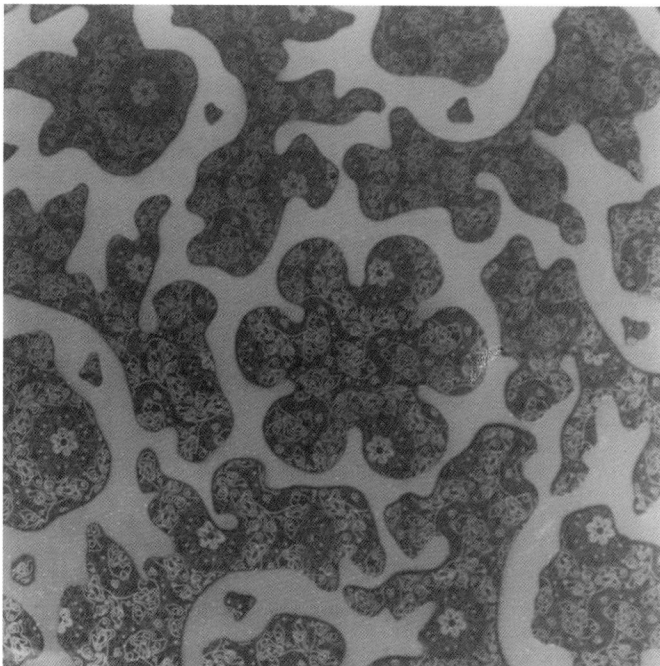

Top left and above, figs.30, 31 *Insulin 8.27 'Oxvar' decorative finish by Vernon Industries. A printed pattern which could be applied to any flat surface, including wood (left) and glass (right).*

fig.32 Design drawing for Pentaerythritol 8.18 ashtray by Wood Brothers.

TECHNICAL INNOVATIONS

The mid 20th century was a period of significant technological advances in the field of new materials. Hartland Thomas was keen to reflect this in the FPG, hence the inclusion of several recently developed synthetic products, including Dunlop's PVC sheeting (FPG 13-14) and Warerite's plastic laminates (FPG 65-70). [fig. 29] The most curious material was Vernon Industries' 'Oxvar' decorative finish, an American patented method of applying printed surface patterns on hardboard, metal or glass panels using a thin hard mastic. [figs. 30, 31]. In addition, Megaw was specifically requested to seek out crystal structures of synthetic materials as a source of decoration; this was why cellulose (the basis for viscose rayon and nitrocellulose-coated leathercloth) was chosen. ICI, one of the largest chemicals firms of the day, was crucial in both respects. Not only did the FPG include their latest products (Mural Rexines and Vynides manufactured by their Leathercloth Division), it also featured diagrams of three substances invented and/or produced by ICI - polythene, nylon and diethyl terephthalate (related to Terylene) – the crystal structures of which had all been researched by ICI scientists.

Most FPG products were surface-decorated, but a few 3-D items were created, notably a pressed glass ashtray by Wood Brothers, decorated with the crystal structure of Pentaerythritol (FPG 80), which was used extensively in the Regatta Restaurant. [fig. 32] Repeated attempts to produce crystal structure tableware, glassware and cutlery for the restaurant proved abortive, however, mainly because of labour and materials shortages affecting these industries. Prototype designs for plates, wineglasses and cutlery were displayed at the Festival, but none entered production. Furniture was also problematic. Originally Hartland Thomas had wanted to collaborate with Furniture Industries Ltd., (Ercol), but their experiments with 'raised contour patterns' came to nothing. Gordon Russell Ltd. were asked if they might use a router to cut patterns in veneer, but declined due to lack of time. Product designer R.D. Russell advised that crystal structure decoration was inappropriate for wood, but suggested tooled leather instead. This idea was finally implemented by Goodearl Brothers (the main restaurant furniture contractors), who produced a bent wood chair with a tooled leather seat back decorated with the structure of Polythene (FPG 25), but only as a one-off.

Architects are notorious for their distrust of pattern design so it is perhaps hardly surprising that aesthetic objections from that quarter hindered the development of certain FPG products. One South Bank exhibition designer, Patience Clifford, initially rejected GEC's Beryl wall lamps (FPG 17) with their lace-covered lampshades, saying: 'my architectural gorge rises slightly at the idea of structural lace!'[37] Misha Black was also rather dismissive at times and offended several FPG manufacturers because of the cavalier way in which he treated them. The Linoleum Manufacturing Company wanted to pull out of the scheme altogether after their stencil-inlay linoleum was rejected for the restaurant. Hartland Thomas persuaded them to remain on board, but they insisted on anonymity.

fig.33
Regatta Restaurant: Hydrargillite 8.33 motif at entrance.

REGATTA RESTAURANT

Finally, in spite of all the ups and downs, the Festival Pattern Group came to fruition. [fig.33] At the press launch in the Regatta Restaurant on 17 April 1951 FPG members proudly sported their Haemoglobin ties (FPG 61). Although, in the end, the interior was not quite as comprehensively crystallographic as originally conceived, crystal structure patterns were omnipresent – on floors, walls, windows and tables. Meanwhile, in the foyer a special introductory FPG display explained the ideas behind the project and presented samples by each manufacturer. [fig.

34] After the launch COID Chairman Gordon Russell heartily congratulated Hartland Thomas on his success in galvanising and uniting such a diverse group of manufacturers: 'I was much impressed by the unanimity of opinion at the Pattern Group meeting. It is obvious that from the Council's point of view you have done a useful piece of work.'[38]

In the restaurant the most dramatic feature was a room divider made of Warner's 'Surrey' furnishing fabric, based on one of Megaw's diagrams, afwillite (FPG 71). [fig. 35] Designed by Marianne Straub, its bold linear contour pattern was intensified by the textile's huge repeat and richly textural yarns in dark green and gold. Wallpapers were used to striking effect as accent features, one decorated with a smaller scale version of Afwillite (FPG 35) in complementary colours, the other featuring Insulin 8.27 (FPG 38), the prettiest and most intricate of Dorothy Hodgkin's diagrams. [figs. 36, 37]. Templeton's Resorcinol carpet (FPG 56), a geometric ball-and-spoke structure, covered the floors, specified in two different colourways: green/purple and purple madder/attic rose. Blackcurrant was the colour chosen for Warerite's Haemoglobin laminate (FPG 68), used as cladding in the coat recesses and on service doors. The same pattern recurred in a woven fabric by Barlow & Jones, used as an alcove curtain, while A.C. Gill's Beryl lace (FPG 19) was used as casements on a large six-bay window. The waitresses wore collars made of Hydrargillite lace (FPG 24), and the same star-shaped motif was embossed on the menu covers (FPG 42) and emblazoned on the entrance wall. Even in the toilets there was no escape from crystal structures: Apophyllite glass screens were incorporated in the ladies lavatories (FPG 8).

fig.34 Festival Pattern Group display in foyer of Regatta Restaurant, featuring samples by all the participants.

Top, fig.35 *Regatta Restaurant: Resorcinol 8.17 carpet by Templeton and Afwillite 8.44 curtains by Warner in foreground; Haemoglobin 8.26 curtains by Barlow & Jones and Insulin 8.27 wallpaper by John Line in alcove behind.*
Bottom, fig.36 *Regatta Restaurant: Afwillite 8.45 wallpaper by John Line; Resorcinol 8.17 carpet by Templeton and Beryl 8.9 lace curtains by A.C. Gill (windows on right).*

Top, fig.37 *Regatta Restaurant: Insulin 8.27 wallpaper by John Line; Haemoglobin 8.26 curtains by Barlow & Jones; Pentaerythitol 8.18 ashtrays by Wood Brothers on tables.*

Middle, fig.38 *Mural on crystallographic theme by John Tunnard in vestibule of Regatta Restaurant.*

Bottom left, fig.39 *Detail of one section of mural.*

Bottom right, fig.40 *John Tunnard working on section of mural featuring crystals.*

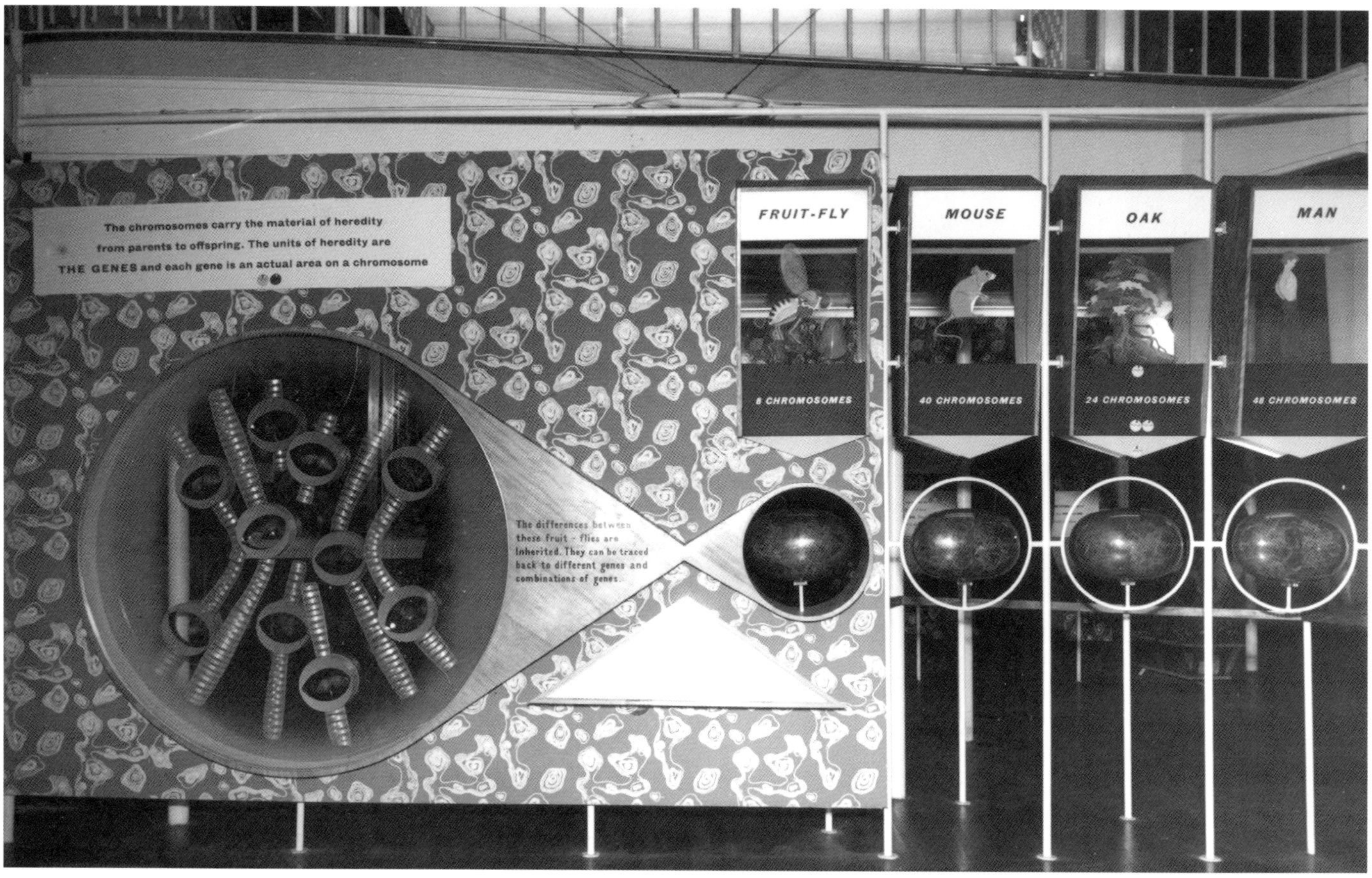

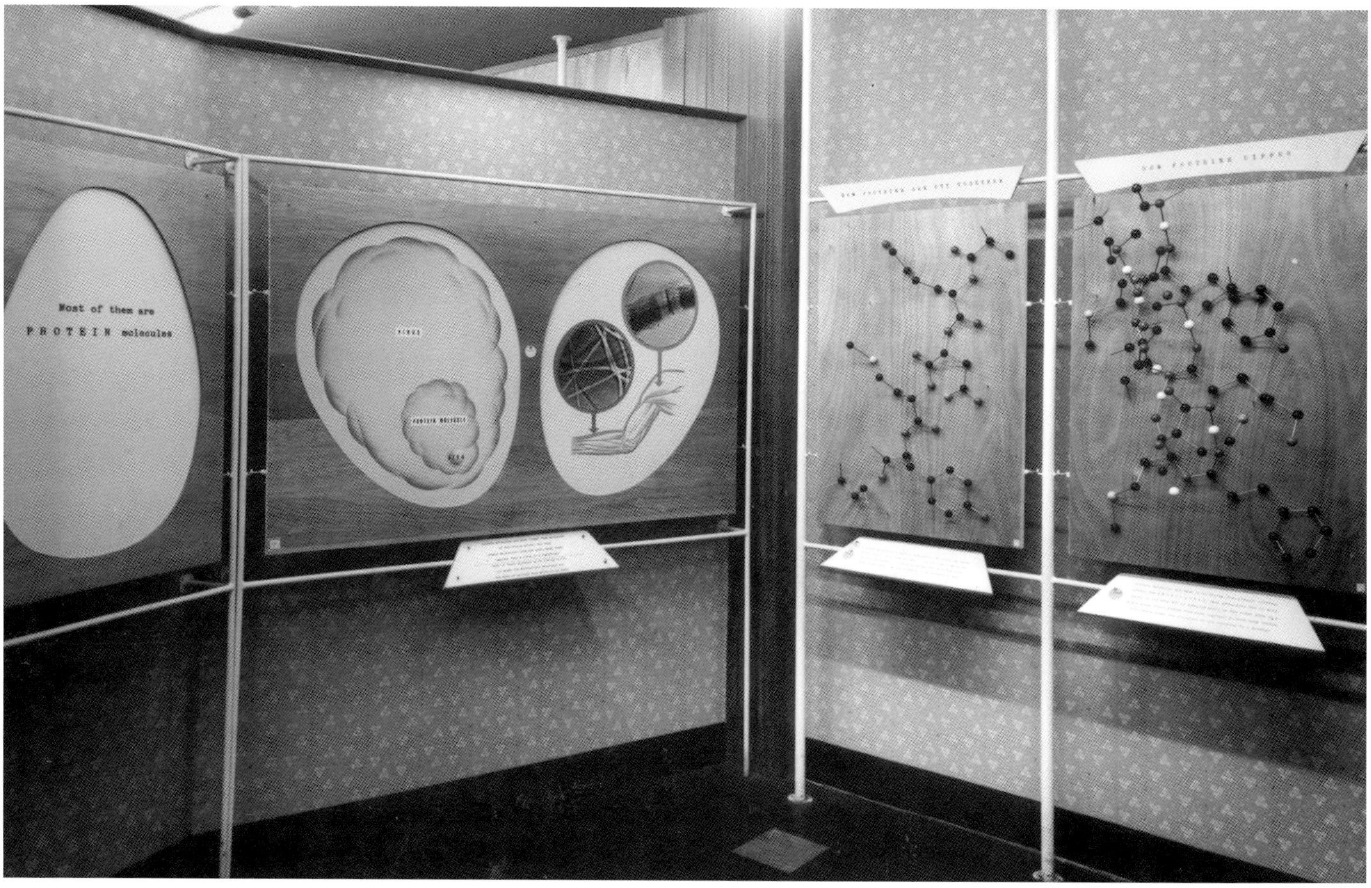

Top, fig.41 *Exhibition of Science: Afwillite 8.45 wallpaper by John Line used as backdrop for display about chromosomes.*
Bottom, fig.42 *Stop Press section, Exhibition of Science: Insulin 8.25 mural rexine by ICI Leathercloth used as backdrop to display about The Structure of Proteins in The Problem of Life (interpretation by Max Perutz).*

fig.43 *Exhibition of Science: Afwillite 8.45 wallpaper by John Line used as backdrop to the displays. 600ft ball-and-spoke light fitting based on the atomic structure of carbon, designed by Brian Peake, produced by GEC.*

Complementing the work of the Festival Pattern Group was an impressive crystallography-inspired mural by John Tunnard, which occupied an entire wall in the vestibule. [fig. 38] With his deep-rooted interest in science, Tunnard was the ideal choice for this commission. Before embarking on the mural he was sent copies of crystal structure diagrams so that he could tie in with the crystallography theme. The finished work featured 'crystals all over the place'[39] – jagged crystalline minerals and snow crystals as well as abstract crystallographic motifs. [figs. 39, 40]

EXHIBITION OF SCIENCE AND DOME OF DISCOVERY

Brian Peake, coordinating designer of the Exhibition of Science, eagerly embraced the concept of crystallographic design. Many screens in the exhibition were adorned with either Afwillite wallpaper (FPG 35) or Insulin leathercloth (FPG 28), providing a stimulating backdrop to the displays. [figs. 41, 42] The exhibition hall itself was dominated by a long winding lighting installation inspired by ball-and-spoke models. [fig. 43] Although not an FPG product, it was clearly designed to complement the crystal structure theme and was produced by one of the group's members, GEC. In the buffet, the walls were adorned with more Insulin Mural Rexine, while the vinyl tiled-tiled floor was inspired by Mica (FPG 15). [fig. 44] The cinema provided another showcase for a cluster of FPG products: the seats were upholstered in Insulin Vynide [fig. 45], while Myoglobin (FPG 32) provided a lively covering for the benches outside. The foyer was decorated with Insulin wallpaper (FPG 37), complemented by a printed cotton curtain featuring magnified Haemoglobin motifs. (FPG 73) Suspended overhead was by a dynamic 'atomic' neon lamp. [fig. 46]

The displays in the Exhibition of Science drew attention to the significance of X-ray crystallography.[40] One section was devoted exclusively to this and featured diagrams, models and equipment, but crystallography was an underlying theme in other areas as well, including Petrology, Metals and Organic Chemistry. Dorothy Hodgkin's X-ray photograph of penicillin was used as an example of Metallic Bonds; Max Perutz supplied interpretation about proteins in a section called The Problem of Life.

The Dome of Discovery also focused on British achievements in science and technology, both historic and contemporary. Designed by Ralph Tubbs, it evoked a flying saucer and provided a magnificent arena for the displays. [fig. 47] The exhibition was in eight thematic sections with broad titles such as The Land, The Earth, The Sea, The Sky, ranging in subject from civil engineering and geology to meteorology and astronomy. [figs. 48, 49] Crystallography was explored in the final section, The Physical World, along with nuclear physics. Its theme, 'how matter – the substance of all things – is made', was divided into two parts, 'how matter is built up (chemistry) or why it behaves as it does (physics)'.[41] A scientific explanation of crystal structure diagrams underlined British achievements in crystallography. This was complemented by a display of products by nine FPG manufacturers: Barlow & Jones, British Celanese, Chance Brothers, Dunlop, ICI Leathercloth, John Line, Old Bleach Linen Company, Vanners & Fennell and Vernon Industries.[42]

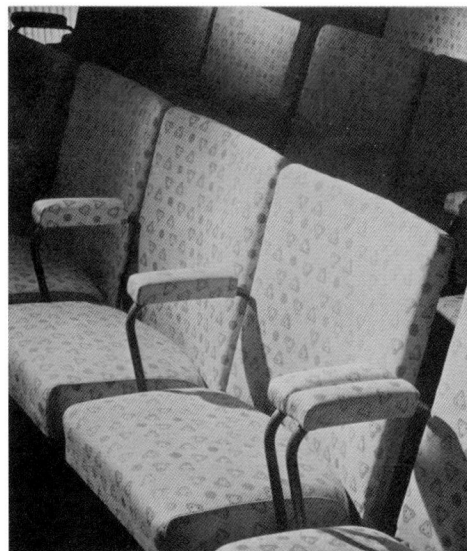

Top, fig.44 *Buffet, Exhibition of Science: Insulin 8.25 mural rexine by ICI Leathercloth on walls (far left); Mica 8.35 vinyl flooring by Dunlop.*

Above, fig.45 *Cinema seating, Exhibition of Science: Insulin 8.25 vynide by ICI Leathercloth used as upholstery.*

Right, fig.46 *Cinema foyer, Exhibition of Science: Insulin 8.25 wallpaper by John Line and Myoglobin 8.46g vynide upholstery by ICI Leathercloth.*

Top, fig.47 *Dome of Discovery, designed by Ralph Tubbs.*

Middle left, fig.48 *Outer Space section, Dome of Discovery, designed by Austin Frazer and Eric Towell, Design Research Unit.*

Middle right, fig.49 *Interior of Dome of Discovery.*

Left, fig 50 *Festival Patterns section of Land Travelling Exhibition.*

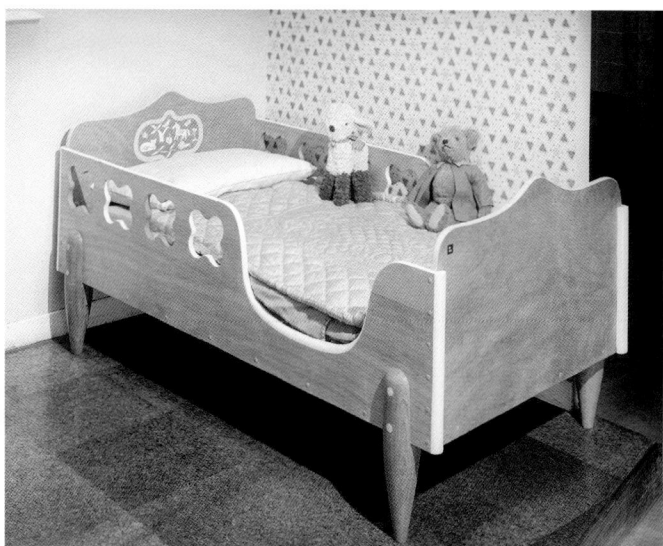

fig.52 Room for a two-year old, Homes and Gardens Pavilion: Insulin 8.25 mural rexine wallcovering by ICI Leathercloth.

fig.51 Exhibition of Science: Afwillite 8.45 wallpaper by John Line used as a backdrop to displays.

The Land Travelling Exhibition, designed by Richard Levin, also incorporated a special section devoted to 'Festival Patterns'. Touring to Manchester, Birmingham, Leeds and Nottingham, it featured a varied assortment of FPG samples, presented in an explosive format, as if the objects were literally bursting out of the wall. [fig. 50]

IMPACT AND MEDIA COVERAGE

The Festival of Britain proved phenomenally successful, attracting 8.5 million visitors over five months from 3 May - 30 September 1951. However, because FPG products were scattered across several locations, their impact may have been somewhat diluted. [figs. 51, 52] The full significance of the FPG may also have been obscured to some degree by the sheer intensity and visual drama of the Festival itself. In the end, there was so much for visitors to see and absorb that even designs as unusual as these were in danger of getting lost. Nevertheless, the event still provided a high-profile showcase for the Festival Pattern Group's work. For most visitors, it would have provided their first introduction to crystallography and their first exposure to contemporary design.

Some impression of the FPG's impact can be gleaned from media coverage. *The Times* declared: 'It is a pleasant change to dissociate the atom from the idea of a destructive bomb and to apply it to the creation of things of beauty.' They praised the collection as 'truly delightful... a refreshing change from

outworn traditional design.'[43] At a local level, too, there was pride in the FPG's achievements. *The East Anglian Daily Times* ran an article headed 'The "Atomic Tie"', referring to Suffolk-based Vanners & Fennell, portentously subtitled, 'A Sudbury "Top Secret" Out – Scientific Triumph for the Festival'.[44] The FPG also benefited from the novel medium of television. Shortly after the Festival opened, Hartland Thomas alerted participants to a programme about the Festival Pattern Group on 21 May 1951: 'Although, as members will be aware, the BBC does not allow manufacturers' names to be mentioned, I am doing my best to include at least one item from each manufacturer member, within the limits imposed by the medium.'[45]

The FPG even attracted the attention of the satirists (always a sign of success). An elaborate spoof appeared in *Punch* entitled 'Petri's Kaleidoscope', in which Hartland Thomas was transformed into M. Pentland-Davis, head of a new research station specialising in 'bacteriographical design'. The author, Bernard Hollowood, mischievously transposes crystallography into microbiology - perhaps an allusion to Perutz's experimental unit (Molecular Structure of Biological Systems). Pentland-Davis is heard lamenting the difficulty of winning public acceptance for the new designs: 'We did a very nice off-white and rose version of the anaerobic bacterium Clostridium Pastorianum, for the cotton people. *That* didn't go. We evolved another pattern from assorted soil bacteria for the potters, and *that* flopped... Gradually though – and this is the point – we were breaking down public resistance to the new art form. Industrialists were getting used to the idea of designs based on bugs.'[46] Helen Megaw retained a copy of this article, so presumably she appreciated the joke.

RESPONSE FROM CRYSTALLOGRAPHIC COMMUNITY

The response from crystallographers was overwhelmingly positive. Megaw reported a few days after the opening that she was 'besieged by inquiries from colleagues and acquaintances to know what is... being made, where it can be had, and at what price.'[47] Sir Lawrence Bragg wrote to Hartland Thomas to say how thrilled he was with the results: 'When in 1922 I worked out the first crystal of any complexity that had been analysed, aragonite, I remember well how excited my wife was with the pattern I showed her as a motif for a piece of embroidery. Ever since then, especially when I was in Manchester, I have been urging industrial friends to use these patterns as a source of inspiration, and I was delighted when Miss Megaw... told me some two years ago that she had aroused your interest. The patterns she showed me yesterday are the practical realization of what we have long wished to see.'[48] Hartland Thomas was so gratified by Bragg's reaction that he asked permission to quote him in the forthcoming *Souvenir Book*, adding: 'The fact that the proposal to use these patterns as a source for decoration came from the scientific side, and has been taken up by artists working in industry, is a happy augury for the rebirth of an understanding between scientists and artists which many of us are hoping to see.'[49]

Bragg's pride in the project is underlined by the fact that, in September 1951, he borrowed a group of FPG samples to show at a public lecture in Stoke-on-Trent. His wife, Lady Alice Bragg, was evidently pleased with the results as well. At the International Congress of Crystallography in Stockholm (27 June-3 July 1951) she wore a spectacular evening dress made of Beryl lace (FPG 19), one of her husband's crystal structures.[50] [figs. 53, 54] Max Perutz (himself the son of a textile manufacturer) was evidently approving as well. His wife, Gisela Perutz, wore a printed rayon dress featuring his diagram Horse Methaemoglobin (FPG 6) at the Stockholm conference. Helen Megaw went to great lengths to solicit samples of FPG products from each of the manufacturers. At the conference she sported a blouse made of Afwillite crepe silk (FPG 33). 'A lot of the men from Cambridge have been wearing the ties,' she reported to Hartland Thomas; 'also someone had sent a tie with

a Patterson diagram of insulin to Dr Patterson, the American inventor of that type of diagram, and he is wearing it with pleasure.'[51] Vanners & Fennell's ties were a big hit, particularly amongst crystallographers. [figs. 55, 56] Nine years later Megaw persuaded the company to reissue their design for China Clay (FPG 59) so that ties could be made for the IUC conference in Cambridge (15-24 August 1960). She herself also had a jacket made from this fabric in 1954.[52]

After all their hard work in bringing the project to fruition, it must have been gratifying to the two chief protagonists, Hartland Thomas and Megaw, to receive such enthusiastic feedback. Hartland Thomas observed at the time: 'Though the chief idea of the Festival Pattern Group was to get leading manufacturers

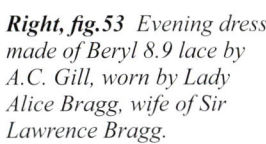

Right, fig.53 Evening dress made of Beryl 8.9 lace by A.C. Gill, worn by Lady Alice Bragg, wife of Sir Lawrence Bragg.

Far right fig.54 Detail of Beryl 8.9 lace by A.C. Gill.

Top fig.55 China Clay 8.6 and Haemoglobin 8.26 ties by Vanners & Fennell. Haemoglobin 8.26 plastic laminate by Warerite behind.

together on a design project and though we were careful to stress that the crystal structure diagrams could not be a ready-made short cut to good design, I always felt that there was more to it than that – if only to add an emotional apprehension to the intellectual study of scientific patterns. / I had not quite realised that the crystallographers' delight in their patterns expressed an interest in their form that had an importance in scientific understanding.'[53]

Knock-on Effects: Later Crystal Structure Initiatives

After the exhibition opened Hartland Thomas and Megaw were both contacted by other various companies interested in using crystal structure designs. In June 1951 the Kidderminster carpet firm Tomkinsons wrote to Megaw requesting illustrations for potential patterns. Shortly afterwards Hartland Thomas was approached by the Manchester firm, Hogg & Mitchell, who wanted to produce crystallographic ties in jacquard-woven crease-resistant rayon. They selected two diagrams, Insulin 8.24 and Afwillite 8.43, which they planned to put on the market the following year.[54] In August 1951 Megaw was contacted by L.W. Mallabar of Oxvar Ltd, the British subsidiary of the American firm responsible for 'Oxvar' decorative finish. He invited her to act as their scientific consultant and to supply crystal structure diagrams for possible use on their products, both in Britain and the US. Megaw entered into a contract with Oxvar on 4 October 1951, and for the next three years sent regular batches of diagrams. Unfortunately none were ever developed for production due to lack of follow-up in the US; the contract was terminated by mutual consent in March 1955.[55]

Crystal structure patterns enjoyed a reprise in 1954 in the décor of the Cunard liner, *Saxonia*. One of the decorative themes of the interiors was Klondike gold, and Megaw was asked to supply a diagram of the crystal structure of gold for a linoleum floor in 'The Yukon Bar'. The Tourist Drawing Room and Library featured two crystal structure wallcoverings made from plasticised fabric by Sanderson: Unorthoclase (a pun on orthoclase) and Myoglobin. A publicity leaflet also referred to 'delicately patterned lace' on the landings, possibly by A.C. Gill. Elaborating on the inspiration behind these furnishings, the brochure declared: 'The beautiful symmetry of these "atomic patterns" is revealed in the crystal structure diagrams which the scientist prepares to record the way in which the atoms of each material are arranged. Here then is released to the designer a whole new world of basic patterns, infinitely varied.'

For Helen Megaw the Festival Pattern Group had always been an enjoyable diversion - albeit a demanding and time-consuming one – from her career as a crystallographer. She kept in touch with several manufacturers afterwards, periodically offering new diagrams or instigating the reintroduction of the original designs. Her last direct contact was in the run-up to the IUC conference in Cambridge in 1960, the programme for which was printed with a new pattern featuring orthoclase. As a Fellow and Director of Studies in Physical Science at Girton College, Megaw was keen to acquire some FPG furnishings for the institution. In October 1951 she arranged for Old Bleach Linen Company to produce some woven cotton fabric for the Lawrence Room at the College featuring the crystal structure of China Clay (FPG 46).[56] Megaw remained at the Cavendish Laboratory until her retirement in 1972. The Apophyllite glass

fig.56 Aluminium Hydroxide 8.4 woven silk tie by Vanners & Fennell.

(FPG 8) panels that she installed in two kitchen doors in her home in Cambridge remains in situ today.

Culmination and Assessment

On 16 October 1951 Mark Hartland Thomas called a meeting to discuss the future of the Festival Pattern Group. The main issues were who would take over from the COID as coordinator, and whether the FPG should continue to focus on crystal structures or broaden its scope. Nine manufacturers were in favour of continuing (Carter & Co., Chance Brothers, A.C. Gill, G.A. Harvey, John Line, Old Bleach Linen Company, Spicers, Warerite and Wood Brothers), eight were unwilling, and the rest were uncertain. In the end it was decided that the FPG had served its purpose and would not be viable in the longer term, so it was better if the group was wound up.[57] Hartland Thomas was later awarded an OBE in recognition of his services during the Festival. He left the COID shortly afterwards to resume his career in architecture. Following his death in 1973, Gordon Russell described him as 'a loyal, hard-working colleague, not easily flustered, who stood up cheerfully to considerable battering.'[58] In terms of his achievements at the COID, the Festival Pattern Group was the pinnacle of his career.

Fortunately for posterity Hartland Thomas wrote a detailed and well-illustrated account of the Festival Pattern Group in the May-June 1951 issue of *Design*, subsequently reproduced as *The*

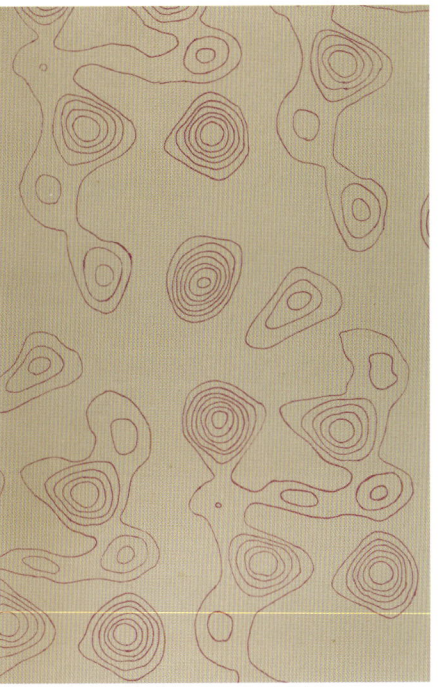

fig.57 Afwillite 8.44 woven fabric ('Surrey') by Warner; Resorcinol 8.17 carpet by Templeton.

fig.59 Afwillite 8.45 diagram by Helen Megaw on the cover of Architectural Review, April 1951.

fig.60 Haemoglobin 8.26 lace (above) and Polythene 8.59c lace (below) by A.C. Gill.

fig.58 Chemical Structure section, Exhibition of Science:.Afwillite 8.45 wallpaper by John Line.

Souvenir Book of Crystal Designs (see Chapter 2). Helen Megaw also published an article at the time called 'The Investigation of Crystal Structure', a much more technical piece explaining the science of crystallography, published in the *Architectural Review* in April 1951. [fig. 59] Both articles emphasise the closeness of the patterns to the original diagrams by illustrating the crystal structures next to the resulting designs.

Throughout the project the scientific significance of the diagrams was constantly reiterated, and the accuracy of the interpretations carefully monitored. When the material was finally presented to the public, it was this aspect of the initiative – its scientific veracity - that was repeatedly brought to the fore. Although some manufacturers' interpretations were freer than others, what is remarkable is how respectfully the source material

was treated. When deviations occurred, Megaw could be highly critical, damning even: 'The patterns I sent to the manufacturers were scientifically correct,' she later asserted. 'But not all the manufacturers kept to these rules.'[59] Retrospectively, it is easy to mistake the literalness of some of their interpretations for lack of imagination, but to criticise the designs for not digressing further from the diagrams misses the point of the scheme – this is exactly what they were *not* supposed to do.

Close attention to detail and an intense engagement with their subject were vital attributes for post-war crystallographers, who often devoted many years to working out particular crystal structures. Converting these diagrams into products that respected these scientific endeavours required sensitivity, application and skill. All the designs had to be transposed from one medium (a dyeline drawing) to another, be it through print, weave, embroidery or mould. This required technical ingenuity, precision and understated creative flair. To appreciate fully the efforts of the FPG - and to understand the motivations behind the project - requires a willingness to engage with the finer detail, both of the scientific source material and the subtleties of the designs.

The huge pattern and rugged contours of Warner's 'Surrey' furnishing fabric (FPG 71) are easy to appreciate for their direct sensory appeal. [fig. 57] The arresting compositions and confident execution of John Line's screen-printed wallpapers also have great power and immediacy. [fig. 58] But other FPG products, although equally heroic, are much more subtle and require much closer scrutiny. The raised threads on the surface of one of Dobroyd's Mica woven wool dress fabrics (FPG 10) cleverly evoke the criss-crossing lines on the original ball-and-spoke diagram. Vanners & Fennell's jacquard-woven tie silks convey the graphic fineness and precision of the scientific diagrams on an appropriately microscopic scale. [fig. 56] A.C. Gill's machine-embroidered lace captures the intricacy and

Above, fig.61 *Advert for Warner & Sons from Design magazine, May-June 1951.*
Above right, fig.62 *Nylon 8.54c woven cotton furnishing fabric ('Helmsley') by Warner, showing five different colourways.*

delicacy of the original line drawings and embodies the idea of a network of connecting threads. [fig. 60] Arguably these miniaturised designs are some of the most successful, both visually and scientifically. Helen Megaw, who could be a harsh critic, seems to have agreed with this assessment: 'On the whole, the lace manufacturers and the makers of tie silks were good,' she concluded four decades on.[60]

COMMERCIAL SUCCESSES AND MANUFACTURING CONSTRAINTS

Although only a small proportion of FPG products went into production, some met with considerable success, notably Chance's Apophyllite glass (FPG 8), marketed under the name 'Festival'. 'As regards reaction to our own Festival Pattern, this has been remarkably good both at home and overseas,' declared J.W. Chance in October 1951. 'Architects have in most cases expressed unqualified approval and the Glass Merchants have not received this new pattern with their traditional gloom. Indeed quite a number of Merchants are enthusiastic about it and good orders have been placed.'[61] Vanners & Fennell's ties were also successful, selling particularly well on the American market. Although their Sales Director, Bernard Rowland, was sceptical about their viability early on because of their high retail cost, sales must have exceeded expectations as the company later came back to Megaw with a request for further designs in 1954.

Other products appear to have been moderately successful within their own sphere, although some by their very nature were intrinsically short-lived – fashion fabrics in particular, which were only ever intended for a single season. Spicers' papers were inherently ephemeral, literally made to be thrown away, as the company themselves admitted: 'Fancy wrapping papers are at the best a fleeting flash of colour and pleasure immediately discarded.' Yet they nevertheless derived great satisfaction from their participation in the scheme: 'Frankly we are very keen on good design and we regard membership of the Festival group as reward in itself. It gave us a chance of working within the scheme of integrated design and for this we are very grateful.'[62]

Some manufacturers took the initiative to develop other crystal structure designs subsequently, another indication of their commitment to the scheme. John Line's 'Cherwell' wallpaper (FPG 40), for example, is clearly based on the structure of Nylon 8.54c, but was not part of the initial FPG range. A.C. Gill continued producing their FPG lace designs until the end of the decade. In 1959 two additional crystal structures, Oxalic Acid Dihydrate and Myoglobin, were mentioned, neither of which were part of the original collection. After a tentative start, Warner & Sons appear to have put several designs into production. Initially only 'Surrey' and 'Rings' were available, but 'Helmsley' was later manufactured as well.[63] [figs. 61, 62] Warerite too, were rather slow to get going. One commentator remarked in 1955: 'It certainly seemed as if the Warerite patterns produced for the Festival Pattern Group were released very reluctantly to a more general market, but I note that they are at last going to put them out more confidently.'[64]

Those FPG products that never progressed beyond prototypes seem to have been hindered by external constraints, mainly resulting from knock-on effects of the war. Cutlery manufacturers, Elkington, were plagued by shortages of essential metals (copper, zinc and nickel); while the pottery industry was hampered by the fact that ceramic litho prints had to be ordered 18 months in advance. Several firms encountered unforeseen technical problems, particularly those working with new materials: Dunlop's PVC sheeting and ICI's leathercloth were both beset by printing hitches, which delayed their original plans to mass-produce these products, although ICI's 1959 Vynide range included an Insulin 8.24 design.

Fresh fields for design...

For modern homes, a design as modern as to-day! Inspired by the atomic structure of aluminium hydroxide. Woven in durable, colour-fast Irish linen. One of the many captivating and unusual designs, originated by Old Bleach for hangings, loose covers and uphols-tery fabrics in linen, cotton, wool or rayon.

Old Bleach FURNISHINGS LTD

Randalstown, N. Ireland

fig.63 Advert for Old Bleach Linen Company from Design magazine, May-June 1951.

fig.64 Interior of Dome of Discovery.

For other companies, such as Templeton and Old Bleach Linen Company, the main obstacle was marketing; in spite of promoting their FPG patterns through advertisements and trade shows, orders were not forthcoming. [fig. 63] Perhaps this was hardly surprising though, in view of the continuing restrictions on the sale of furnishings on the home market in 1951; exports were still the primary focus at this date. Another factor to bear in mind is that these designs were trailblazers: like the Dome of Discovery and Skylon, they must have looked incredibly futuristic in 1951. [figs. 64, 65] It was inevitable that such unconventional designs would face resistance in the

fig.65 Exterior of Dome of Discovery.

marketplace, not so much from consumers - who actually proved quite receptive to radical abstract patterns during the 1950s and actively embraced the notion of 'Contemporary' design - but from the notoriously conservative buyers who controlled (and censored) what was stocked in the shops.

After all the razzmatazz of the Festival opening, everything went rather quiet on the Festival Pattern Group front. Liberty's held a two-week display of FPG furnishings, including carpets, in their Regent Street store in mid August 1951, but as time went on Megaw became increasingly frustrated at what she perceived as lack of effort by FPG manufacturers to capitalise on the project and get their products into the shops. 'A long sad tale of opportunities lost after such a promising start,' is how she characterised the situation retrospectively.[65] However, she may not have appreciated the practical difficulties faced by manufacturers in adapting from wartime to peacetime production, and in re-establishing their shattered manufacturing and trading infrastructure after a decade of disruption. In 1951 the economy was in a dire state and the domestic market was in a state of flux. Most FPG manufacturers were simply not geared up to cope with the expectations that the Festival generated.

LEGACY

Although the public and the media found the FPG intriguing from a scientific point of view, the concept was too specific – and too idiosyncratic - to lead to a new school of crystal structure design. Instead, pattern developed its own momentum during the 1950s, driven forward by the talents of free-thinking abstract designers such as Lucienne Day. The legacy of the FPG can be discerned in the vogue for spindly linear patterns evoking micro-organisms and molecular forms, popular on 1950s textiles. Ripple effects can also be gleaned from the title and imagery of a few specific products later in the decade, such as Tibor Reich's 'Atomic' printed furnishing fabrics (1960) and Midwinter's 'Festival' tableware by Jessie Tait (1955), the latter loosely evoking crystallographic motifs. Direct parallels are found in a printed furnishing fabric called 'Atomics' designed by the Swedish nuclear scientist The Svedberg (1884-1971) for the 'Signed Textiles' collection produced by Nordiska Kompaniet in 1954, featuring electron configurations and the deflections of charged particles in magnetic fields.

fig.66 Advert for Mullard from Guide to Exhibition of Science.

fig.67 Decorative screen, Waterloo Road, Festival of Britain, designed by Edward Mills.

The fascination with all things atomic, particularly three-dimensional atomic structure models [fig. 66], led to a craze for accessories with ball-and-spoke structures, such as coat hangers and magazine racks. First manifested at the Festival of Britain in a huge architectural screen featuring suspended multi-coloured balls [fig. 67], it culminated on a grandiose scale in the Atomium at the Brussels World Fair in 1958. Interestingly, this event was exploited by British crystallographers to showcase their recent achievements: the International Science Pavilion featured John Kendrew's initial myoglobin structure model and two versions of Crick and Watson's double helix model of DNA.

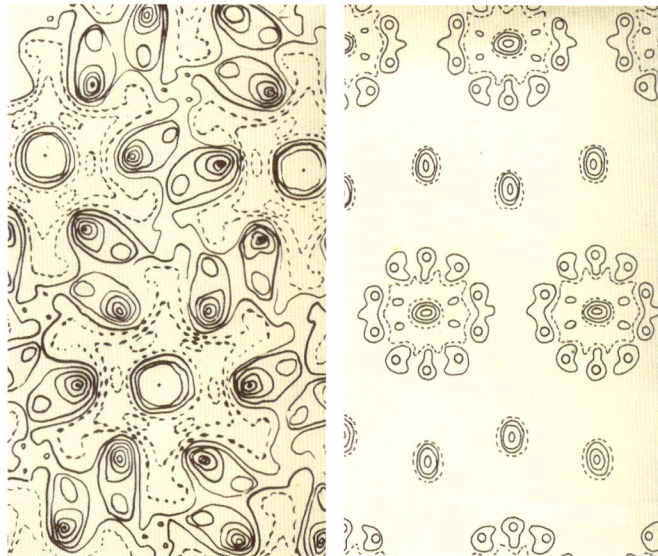

figs. 68-69 Two crystal structure diagrams by Helen Megaw, representing Quinol (left) and Nickel-Manganese-Aluminium alloy (right) from Pattern & Texture by J.A. Dunkin Wedd, 1956.

However, although atomic structure models caught the public imagination and provided inspiration for post-war product design, very few designers engaged with crystallography in a meaningful way. Consultant designer J.A. Dunkin Wedd was unusual in this respect. He wrote to Megaw in 1955 requesting illustrations for a forthcoming book, *Pattern and Texture – Sources for Design*, published the following year.[66] Five pages were devoted to Megaw's crystal structures, all new patterns, mostly of the contour map variety, including Quinol and Nickel-Manganese-Aluminium alloy. [figs. 68, 69] These patterns are highly appealing and their visual interest clearly derives from their unconventional source. Wedd's commentary is somewhat eccentric, but the point he makes is valid: 'generally speaking the aim of the scientist is to find a pattern in events, and this sometimes comes over in a graph or a diagram. What the Festival Pattern Goup demonstrated, and this book reiterates, is the need of some focus to which this wasted material could be directed.'[67]

Retrospectively the Festival Pattern Group appears something of a curiosity, although curiously fascinating nevertheless. Most FPG manufacturers recognised at the time that the project was not primarily a commercial enterprise, but rather an intriguing creative experiment, visually compelling and artistically valid for its own sake. Perhaps this is how it should be judged for posterity. W.A. Dickie, Director of British Celanese, summed up this disinterested attitude in an appreciative letter to Hartland Thomas: 'I must say that the designs have come out much better than I anticipated. They strike an air of novelty and yet contain a unity which is mentally even more than visually perceived. I suppose that, in a way, was the underlying idea.'[68]

THE SOUVENIR BOOK OF

CRYSTAL DESIGNS

The fascinating story in colour of the FESTIVAL PATTERN GROUP

FESTIVAL PATTERN GROUP

by Mark Hartland Thomas, MA, FRIBA

Chief Industrial Officer, Council of Industrial Design

TWENTY-EIGHT LEADING British manufacturers have been working together at the invitation of the Council of Industrial Design for some 18 months upon a programme of design development in connection with the Festival of Britain. Since co-operation on this scale between manufacturers, in the field of design, is of even greater interest than the idea that brought them together, I am relating the story of the Festival Pattern Group at some length, as well as illustrating their work and explaining its theme.

First the theme—then the reader can look at the pictures and come back to the story. The products illustrated in these pages are all decorated with patterns derived from crystal structure diagrams: these are the maps that a scientist draws to record the arrangement of the atoms in particular materials.

A crystal structure diagram takes the form of a repeating symmetrical pattern, like a wallpaper. Examples of them are given below. They vary in character according to the different materials examined, or the different features in a particular material brought out in the diagram (like the difference between a relief map and a road map of Britain), or the difference in plane at which the cross-section has been taken (like the difference between the ground plan and the elevation of a building).

Considerable ingenuity has been required of the industrial designer in adapting a diagram to his own medium, and the designer's name is given with the manufacturer's in the captions on the following pages.

The project began in May 1949, when I attended a week-end course at Ashridge organised by the Society of Industrial Artists. The theme of the course was to show industrial designers some visual material from other arts and sciences in order to broaden our minds. Among the papers read was one by Professor Kathleen Lonsdale on crystallography, in which she remarked that crystal structure diagrams might be used in textile designs. At lunch after her lecture I said I was thinking of taking up her suggestion with manufacturers; but first, where should I get the material? She at once referred me to a colleague, Dr Helen Megaw, of Girton College, Cambridge, who, she said, had drawn out some of the diagrams as a basis for decoration.

I wrote to Dr Megaw and she sent me an article that she had written for publication. A significant point in the article was that whilst in the popular view a scientist laboured at research in order to gain control over the forces of nature, this was often not his real motive. Too many crystallographers the chief incentive was sheer delight in the beauty of the patterns in

EXAMPLES OF CRYSTAL STRUCTURE DIAGRAMS

 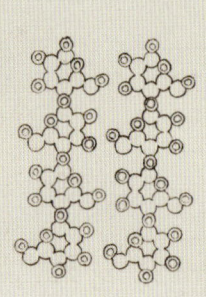 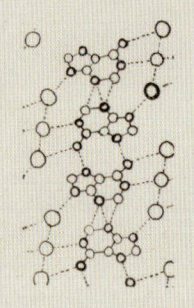  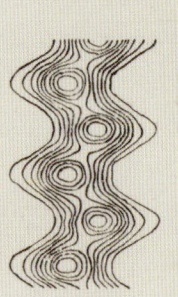 

PEROVSKITE   **CELLULOSE**

The diagrams reveal that perovskite is close-packed and continuous in three dimensions, whilst cellulose is fibrous and stringy

ADENINE HYDROCHLORIDE

Both diagrams above relate to the same material, as can be seen by comparing the unit of structure of the two patterns. The difference in character is due to the former showing by dot and line the positions of the atoms and the forces binding them strongly together, while the latter is a "Patterson map" showing by contours the interatomic distances in the structure

POLYTHENE

Both diagrams relate to polythene but the difference is one of plane of section. The former shows part of one molecule which is indefinitely long: the latter is a cross-section the other way, showing a group of such molecules side by side

From a letter to Mark Hartland Thomas from Professor Sir Lawrence Bragg, F.R.S., Cambridge, 11 May 1951

"... I must write to say how delighted I am.... When in 1922 I worked out the first crystal of any complexity that had been analysed, aragonite, I remember well how excited my wife was with the pattern I showed her as a motif for a piece of embroidery. Ever since then I have been urging industrial friends to use these patterns as a source of inspiration, and I was delighted when Miss Megaw told me two years ago that she had aroused your interest. The patterns she showed me yesterday are the practical realisation of what we have long wished to see...."

Dr Helen Megaw of Girton College, Cambridge, the Festival Pattern Group's scientific consultant, setting an X-ray goniometer to take photographs from which a crystal structure diagram is calculated. In the background a three-dimensional model of afwillite can just be seen. Dr Megaw is also Assistant Director of Research, University of Cambridge, Department of Physics, Cavendish Laboratory, Cambridge

the basic structure of nature that were revealed, often for the first time, by their studies. This opinion, coming from the scientific side, was a happy augury for the project that was taking shape in my mind.

I asked Dr Megaw to agree not to publish her article, explaining that though the proposal to use crystal structure diagrams for textiles would evoke interest, it would be only a passing one: that an idea requiring expensive development for its realisation would be taken up by nobody if it was broadcast to everybody. This principle also made it necessary for me, in order to launch the project, to approach individual manufacturers and to limit the invitations, on a Rotarian system, to one manufacturer for each kind of product. As an acknowledgment, those who accepted made a contribution towards expenses.

continued overleaf

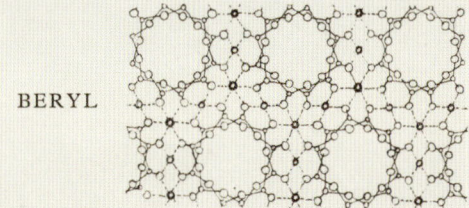

BERYL

Below, wall-bracket light: manufacturer, GEC (designer, R. J. Reynolds). Curtain-lace: manufacturer, A. C. Gill Ltd (designer, H. Webster). $1\frac{1}{2}$ x $2\frac{3}{4}$in. repeat. These pictures epitomise the spirit of co-operation in the Group; the pattern in the shade is the actual lace shown alongside it (which can be seen in the Regatta Restaurant), and the same pattern is echoed in the plastic backplate by a third member, Warerite Ltd

 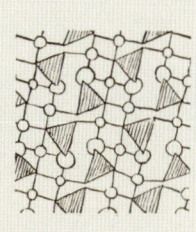 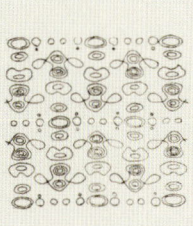

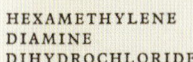
HEXAMETHYLENE DIAMINE DIHYDROCHLORIDE

AFWILLITE

Besides the differences for scientific reasons, there is a wealth of variety in style as the patterns meet the eye and recall quite unscientific associations. The first of these three might almost be Tudor strap ornament, the second cocktail glasses and bubbles behind a bar, the third has a niggling little Paul Klee flavour

 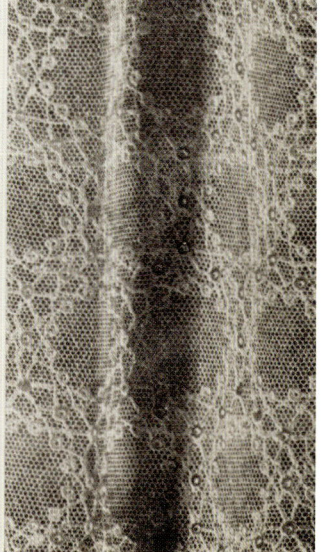

Sketch for a dress print: Arnold Lever (London) Ltd: repeat 15 x 15in. (approx).
The designs on these two pages are derived from the haemoglobin diagram shown opposite. This one, by Arnold Lever, goes furthest from the original in adaptation

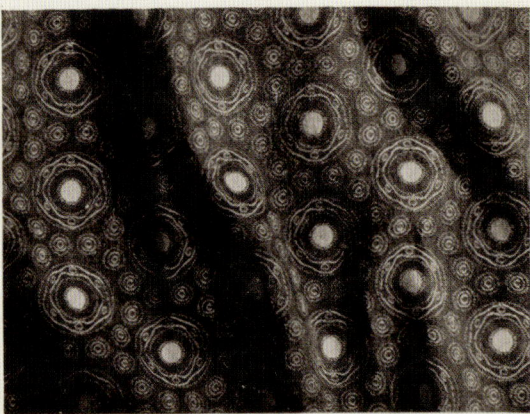

Curtain fabric, woven in yarn-dyed viscose rayon and cotton: Barlow and Jones: repeat 2½ x 4in.
A green version of the same fabric appears on the front cover. This design, by contrast with Lever's dress print, is very close to the original

Screen printed cotton: Warner and Sons: repeat 14 x 26in. This can be seen in the cinema foyer at the Exhibition of Science, South Kensington

Photo-lithographic transfer proofed on dinner plate: Royal College of Art (Peter Wall): repeat 1⅛ x 1⅞in. Screen printed cotton cambric: Barlow and Jones: repeat 2¼ x 3¾in.

Dr Megaw agreed not to publish her ideas for the time being and, later, to be retained as scientific consultant. Her main contribution has been the essential one of supplying the crystal structure diagrams, both from her own researches and from those of scientific colleagues, arranging payment for non-exclusive licences under their copyright. She has also explained the diagrams to members of the Group and checked the accuracy of scientific statements in our written material (including the present article).

Before concluding these arrangements for the continued supply of the diagrams, I had shown some of them to colleagues at the Council of Industrial Design and confirmed my own opinion that here was very promising raw material for applied decoration. I had it in mind that we are at a stage in the history of industrial design when both the public and leading designers have a feeling for more richness in style and decoration, but are somewhat at a loss for inspiration. Traditional patterns that have come down to us from ancient Greece and elsewhere, had lost much of their sparkle by now; and the fashionable alternative of a doodle on a piece of paper, folded for symmetry, could hardly lay the foundations of a new school of design.

But these crystal structure diagrams had the discipline of exact repetitive symmetry; they were above all very pretty and were full of rich variety, yet with a

2

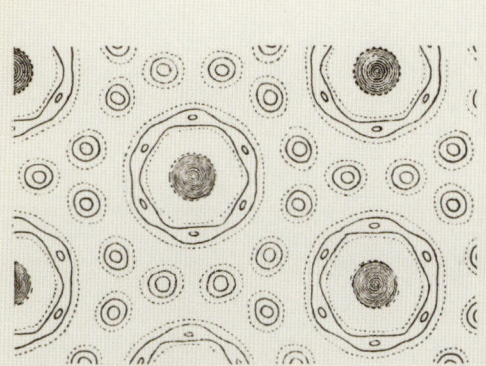

HAEMOGLOBIN

Embroidered cotton lace: A. C. Gill (H. Webster): repeat 2¾ x 3¼in.
The necessities of the machine have required departures from the original that bring out the character of the material

Plastic sheet: Warerite (Martin O. Rowlands): repeat 1⅜ x 1⅜in.
Tie silks: Vanners and Fennell (B. Rowland): repeat ½ x ⅞in. and 1⅜ x 1⅜in.
The left-hand tie is from the china clay diagram, p. 13
The right-hand design is perhaps the most successful adaptation of the haemoglobin diagram, though the triangular framework has been altered to a square

ICI Leathercloth (C. Garnier). repeat 12½ x 9in., 5¼ x 3in., and 1¼ x 2⅛in.
Though it is not the first time ICI leathercloth has been made in pretty colours, these examples and others on p. 7 will be the big surprise in the project for most people. The material is for use in upholstery and wall-covering where heavy wear is anticipated, especially in vehicles. It can be found in several places in the furnishing of the Exhibition of Science, South Kensington

3

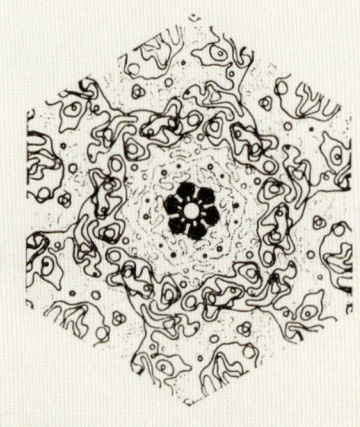

INSULIN

Bride and bridesmaids' veils in lace. A. C. Gill (H. Webster). A free adaptation using part of the diagram as an isolated motif

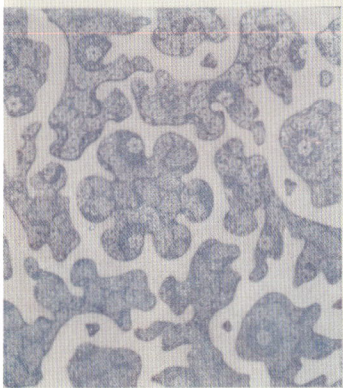

Oxvar *decorative finish: Vernons Industries (W. T. Higgins): repeat 7 x 4in. and 31 x 17in.*
By this offset process a decorated wearing surface can be applied to many different materials in furniture, wall-coverings and the like. The witty device of using the pattern twice, at two different scales, can be seen again in the Warerite sample on p. 12

Wallpaper: John Line and Sons (William J. Odell): repeat 21 x 21in.
The change from the hexagonal grid of the diagram to a square repeat was radical, but despite the bold treatment, character is not lost. To be seen in the Regatta Restaurant
Stencil-inlay linoleum: manufacturer, anonymous (designer, E. H. Tee), repeat 18 x 36in.
The same design as on front cover, with an alternative colouring inset, bottom right. The bright colours are for those who like Turkey carpet effects, without giving them an imitation carpet—from an industry that has produced imitations of every other conceivable floor-covering. The grain here is not imitation carpet-tufting, but the proper result from an ingenious mechanical process

Polyvinyl chloride sheeting: Dunlop (Mary A. Harper): repeat 9½ x 10¾in.

4

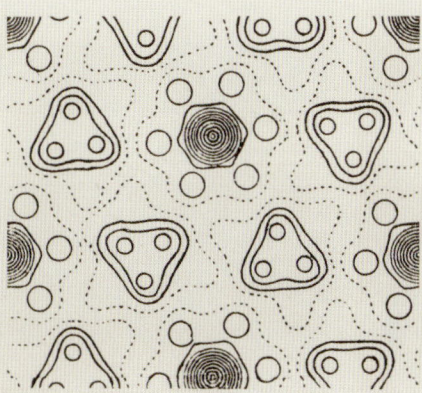

INSULIN

On right, wallpaper: John Line and Sons (Robert Sevant): repeat 10½ x 10½in.
Below, I C I Leathercloth (C. Garnier): repeat 2 x 1⅛in.

Three very different adaptations of the same diagram. Sevant converts the hexagonal repeat into a square one and drops half the original in doing so. Garnier emphasises the motif that Sevant has omitted, whilst Brown lays the emphasis the other way, slightly displacing the subsidiary motif

Below, carpet: James Templeton and Co Ltd (G. Brown): repeat 14½ x 27in

remarkable family likeness; they were essentially modern because the technique that constructed them was quite recent, and yet, like all successful decoration of the past, they derived from nature—although it was nature at a submicroscopic scale not previously revealed.

To check their application to textiles, the branch of industry most likely to be concerned, I consulted Dennis Lennon, then Director of the Rayon Industry Design Centre. He referred to one of his leading industrial members (who later joined the Group) and gave me an encouraging answer. The next move was at the Exhibitions Presentation Panel of the Festival of Britain. I reported to my colleagues there that I had a special project under way for which I wanted to

earmark provisionally one of the restaurants of the South Bank Exhibition. Misha Black volunteered to hold the Regatta Restaurant, for which he was architect, until I was ready to tell them more about the proposal. This was a particularly happy chance for I found later that Dr Megaw's article, mentioned above, had originally been written for a publication in which his firm, Design Research Unit, was concerned. It meant that he and his fellow-architect Alexander Gibson were already sympathetic to the idea; they have worked enthusiastically with the members of the Group in the furnishing and decoration of the Regatta Restaurant as the centrepiece for the promotion of the project.

There was one more check to make before going

5

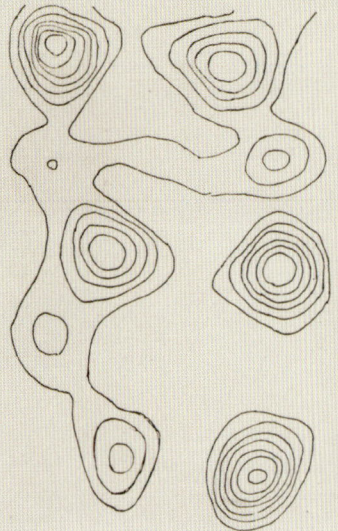

AFWILLITE

Dress print: British Celanese (S. M. Slade): repeat 12 x 6½in. The Celanese examples, here and on p. 7, make play with the contour type of diagram which is like the free-shape outline of current fashion

Plastic sheet: Warerite (Martin O. Rowlands): repeat 2 x 7in. and 4 x 14in. As in the Vernon sample (p. 4), the designer has superimposed the pattern again, to a larger scale

Sketch for warp tapestry curtain: Warner and Sons (Marianne Straub): repeat 22 x 32in.

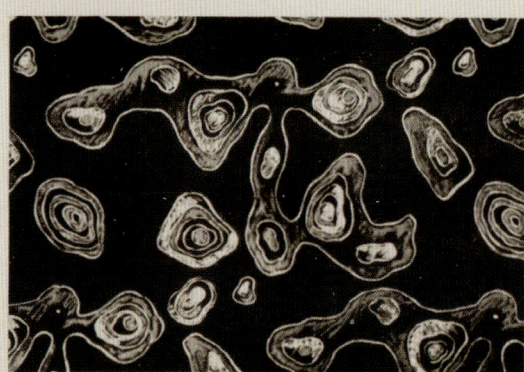

Wallpaper: John Line and Sons (William J. Odell): repeat 21 x 21in.

ahead with it as a Festival of Britain project, namely on the scientific side. I share with Ian Cox (Director, Science, for the Festival) responsibility for the development of the theme of the South Bank and other exhibitions: he takes the scientific side and I the industrial. So I told him how I was planning to steal some of his scientific thunder and apply it to mere industrial art—Jupiter's fire degraded to Vulcan's forge. He exclaimed that this was the best thing that had happened that year, and explained the reason. Crystallography was a branch of science not only of first importance in the modern world, as I knew, but also one that is particularly highly developed in Britain, and it was to figure in the Dome of Discovery and at considerable length in the Exhibition of Science, South Kensington. I was certainly to make the results of my project available to support his displays. This I readily agreed to and results will be seen at the Science Exhibition, South Kensington, where several members of the Group have contributed to the furnishing and equipment. *continued on page 8*

Left: These two can both be seen in the Regatta Restaurant. Compare the difference in scale and treatment of the pattern, done to harmonise together, assisted by matching colours. Difficulties of marketing custom have frustrated attempts to match designs and colours between different manufacturers, but it remains a promising field for design development, especially if matching sets could be sold ready-made

6

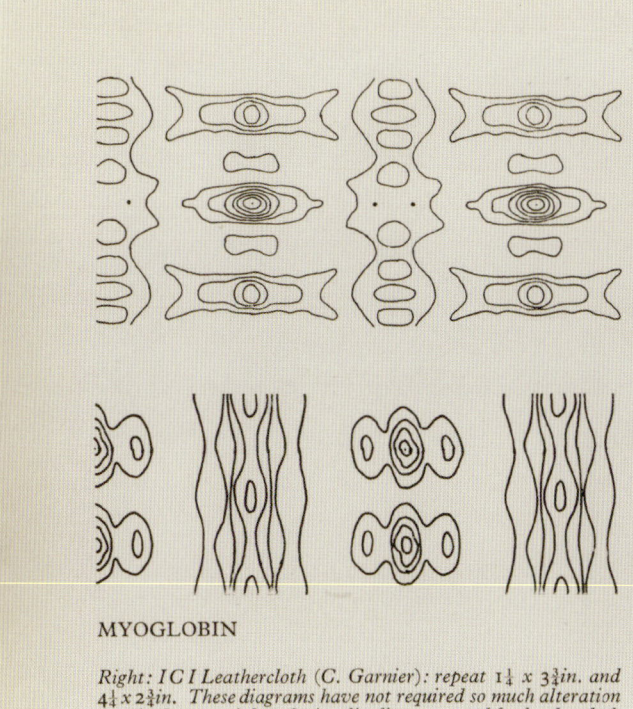

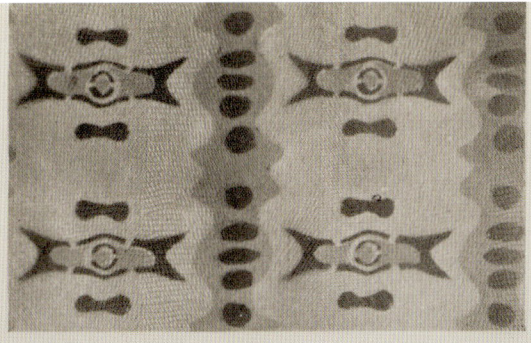

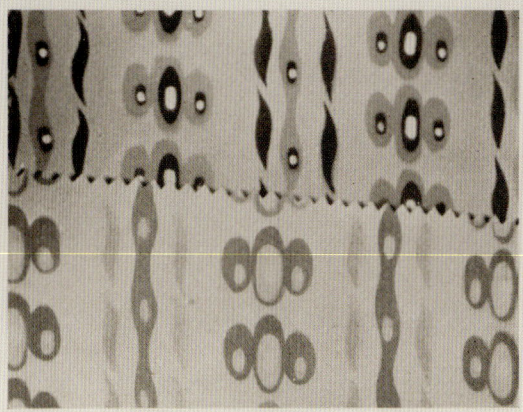

MYOGLOBIN

Right: I C I Leathercloth (C. Garnier): repeat 1¼ x 3¾in. and 4¼ x 2¾in. These diagrams have not required so much alteration to suit the material as the insulin diagrams used for leathercloth on previous pages

HORSE METHAEMO-GLOBIN

Dress print: British Celanese (S. M. Slade): repeat 5⅛ x 3⅝in.

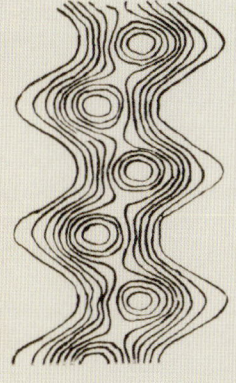

POLYTHENE

POLYTHENE

Embroidered cotton lace: A. C. Gill (H. Webster): repeat 1 x 1½in. Right, cutlery and flatware: Elkington and Co (H. G. Bowring). It is easier to spoil silverware by decoration than to improve it. Here the designer has graduated the pattern in sympathy with the profile and made a decorated handle that is as good as, if not better than, the plain one (p. 13)

7

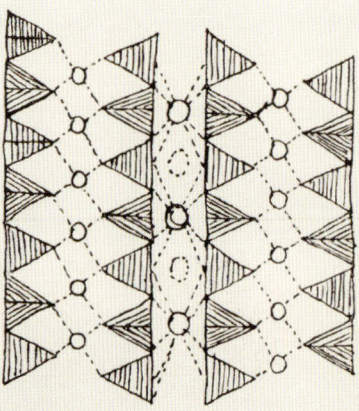

MICA

Woollen dress fabrics: Dobroyd (Tony Dawson): repeat ½ x 2in. and ½ x ⅝in. Two essays on the same diagram. The blue one is also done in bright red and in yellow

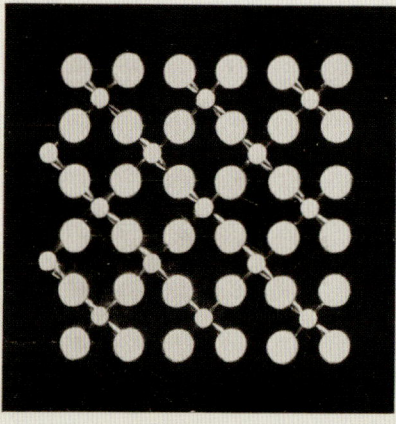

Pierced metal sheet: G. A. Harvey: repeat 1 x 1 inch. Pierced metal has a thousand applications, from ventilator grilles to waste-paper baskets. The piercing, which is required by function, necessitates some sort of pattern. In the past such patterns have been too ornate; nowadays they are perhaps too modest. In the Group's experimental work, Harvey's did several that were more ambitious than this. Public interest in the Festival Pattern Group may encourage them to make tools for some of them

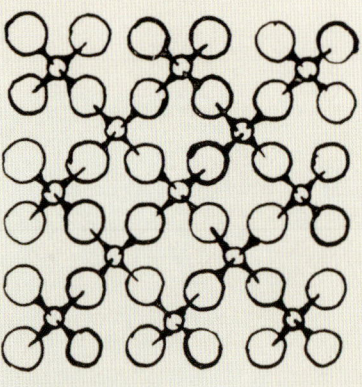

CRISTOBALITE

We agreed that the use of crystal structure diagrams in industrial design was a true development of the theme of the South Bank Exhibition, which was to be a combined exhibition of science, technology and industrial design. Indeed, we reflected that a hundred years ago the Prince Consort had tried to bring art and science together again in harmony, but had failed. We were attempting the same task again at his centenary and this was to be one of the means of carrying it through.

The preliminaries were now completed—by mid-August 1949—and the next four months were spent in collecting members for the Group and helping them to get to work. I was careful to explain that I was not providing a ready-made short cut to good design: the scientific diagrams were only a source of inspiration for designers to use in creative work.

The first meeting of the Group took place on 16 December 1949, by which time 14 manufacturer-members had joined. In deciding to which firms to extend invitations, I had to think which were likely to venture with us in a new and untried idea. Moreover, I wanted most of them to be leading firms of world-wide scope, for I hoped thus to promote not only good design but also successful exporting. The idea was for members not only to develop products for display in the Festival exhibitions, but simultaneously with the opening of the Festival to place other products, similarly decorated, with their agents in foreign markets to take advantage of the world-wide interest that would be directed towards these new designs from the Festival of Britain. Several members of the Group have been able to do this, though perhaps not as many as I had hoped.

8

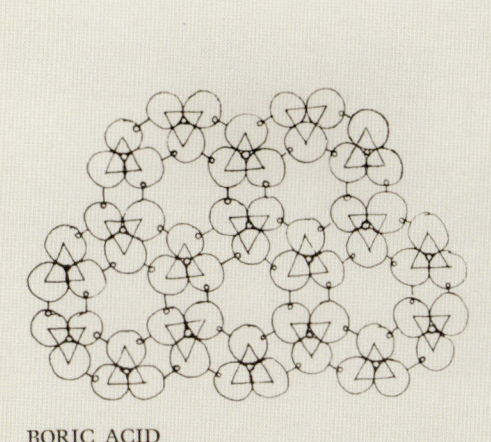

BORIC ACID

Wallpaper: John Line and Sons (William J. Odell): repeat 21 x 18in.

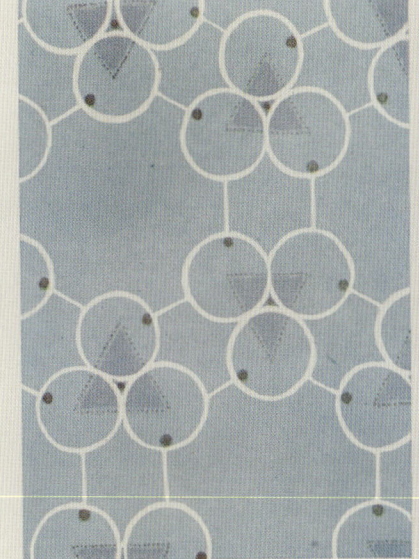

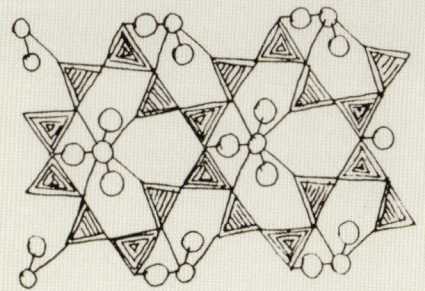

ORTHOCLASE

Furnishing fabrics: The Old Bleach Linen Co: repeat 6¼ x 7¾in. and 6¼ x 3¾in.

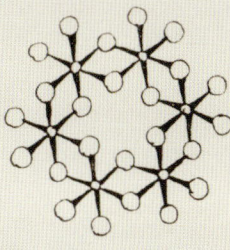

HYDRARGILLITE

Old Bleach and Warerite have both copied the hydrargillite diagram faithfully, but have altered the spacing differently to suit the two different applications. Plastic table top: Warerite (Martin O. Rowlands)

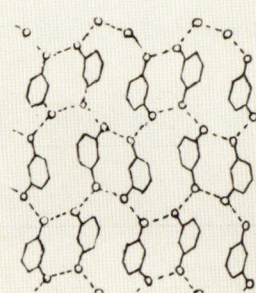

Carpets: James Templeton (R. Anderson): repeat 7 x 4½in. The same design is here shown in two colourings: both to be seen in the Regatta Restaurant. The small dots, representing atoms, have been omitted, and the lozenges have been straightened to suit the weaving

RESORCINOL

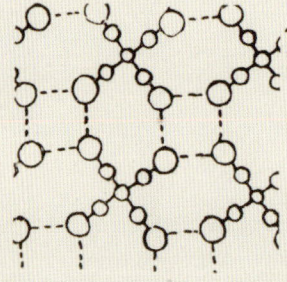

PENTAERYTHRITOL

Moulded glass ashtray: Wood Bros Glass Co (E. Sykes): diameter 4¾in. This ashtray is the only product that attempts to portray the three-dimensional symmetry shown in a crystal-structure diagram—except for some exciting light fittings produced by GEC for the Science Exhibition. The ashtrays are used in the Regatta Restaurant

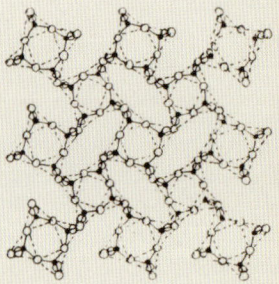

Figured rolled glass: Chance Bros (J. Beresford Evans): repeat ½ x ½in. This design is as unobtrusive as the random-pattern obscured glasses used most frequently today; but it has a delicate charm as well, which comes from the subtlety of the exquisite little design. Two pitfalls in window-glass design were avoided—to prevent cutting to waste for matching, the pattern is on a slant and the repeat small; and so that sunshine does not burn the curtains, the blobs themselves are small

APOPHYLLITE

Although I aimed high, the response was remarkable. In those early days everyone said "Yes." There were no refusals until quite recently, when I have attempted to fill the last few gaps in the coverage of industries that could usefully employ decorative pattern. In recent weeks some have regretfully declined on the ground that there was too short a time before the Festival for them to make an effective contribution.

The spirit at meetings of the Group has been

10

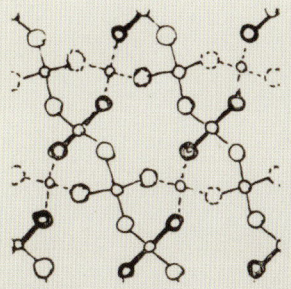

QUARTZ

Carpet: James Templeton (R. Anderson): repeat 8½ x 10½in. One set of dots, for atoms, has been omitted in this adaptation. The diagrams are specially suited to making patterns for carpets, because the strongly marked lattice patterns measure out the space in a room and give it scale

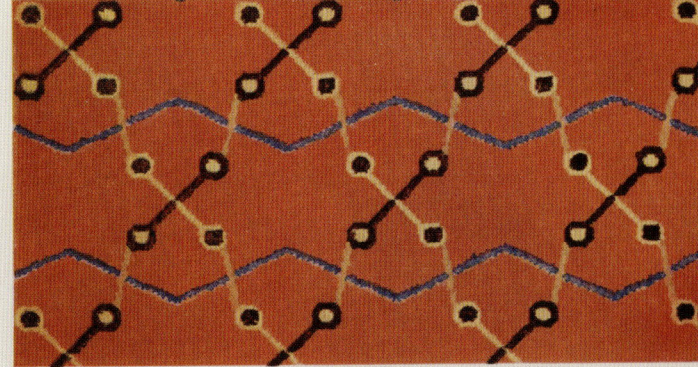

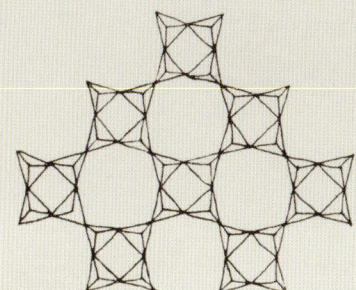

APOPHYLLITE

Plastic sheet: Warerite (Martin O. Rowlands): repeat 2 x 2in.

Below, wall-tile, 3 x 2ft.: Carter and Co (Reginald Till): repeat 10 x 17ft.

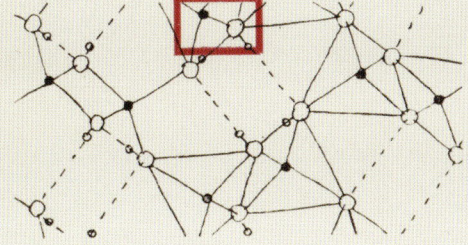

ZINC HYDROXIDE

The portion of the zinc hydroxide diagram included in the sample tile, right, is marked in red on the diagram above. The tile is thus only a small fraction of the design, which is meant for a very large plain wall area

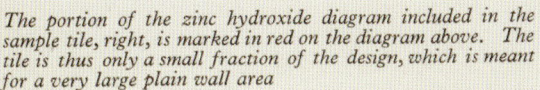

extraordinary. It has been customary for members to bring examples of their work to meetings for discussion and to vie with one another. There has also been an exchange of ideas between meetings. That this work got off to a flying start is particularly due to Hugh McKenna, Templeton's designer, who stole the show at the first meeting with a portfolio full of the most adventurous and stimulating carpet designs.

Apart from the chief importance of the Festival

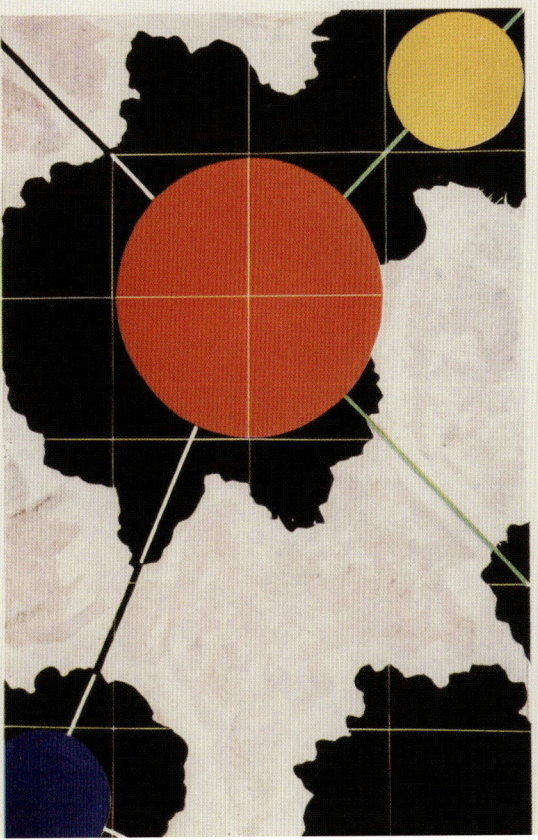

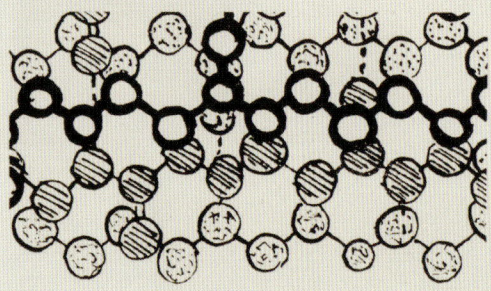

NYLON

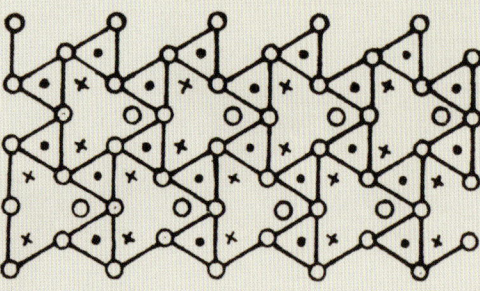

CHINA CLAY

Furnishing fabrics, jacquard woven: Warner and Sons Ltd.
Red and green fabrics (Marianne Straub): repeat 34 x 24in.
and 17 x 12in. Blue and yellow (Alec Hunter): repeat 3 x 4½in.

Pattern Group as an essay in co-operation in design development, readers will judge it mostly by its fruits. These can be appraised in the illustrations to this article and in certain displays at some of the Festival exhibitions.

The centrepiece for display of the patterns is the furnishing and equipment of the Regatta Restaurant. In this the Group has been most fortunate in the appointment of the caterer for the restaurant, for Ernest Corscadden, manager for Hagenbach's of Wakefield, is a man of taste and culture who has entered fully into the spirit of the project.

Not all members of the Group have provided furnishings for the Restaurant, but the public will notice Festival Patterns at a number of points, on the exhibits or on the equipment of the Festival exhibitions. They will see them, too, in shops—abroad as well as at home.

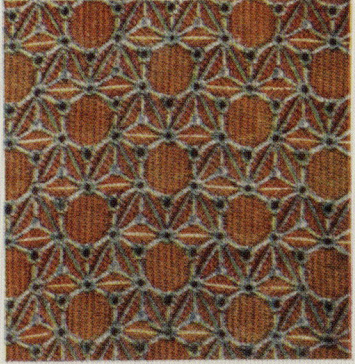

Tie silk: Vanners and Fennell (B. Rowland): repeat ½in. x ⅞in. Here is shown, in actual size, the material of the left-hand tie of the pair shown on p. 3, for comparison with other designs based on this china-clay diagram

Above; linen damask table-mat and napkin: Old Bleach: repeat 1¼ x 2¼in. (The flatware is of the same basic pattern as that shown on p. 7 but without the decoration)

Below: wineglass with enamelled motif: Stevens and Williams (S. W. Thompson): motif 1 x 1in.
Part of the same diagram as for the textiles shown on this page has been extracted and modified as an isolated motif

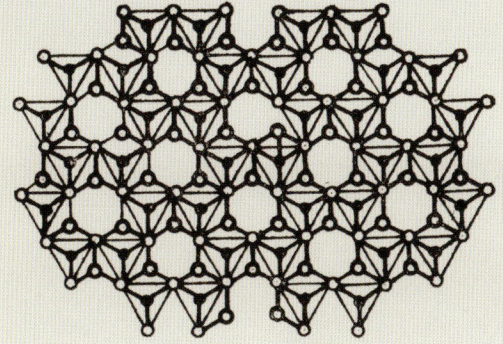

CHINA CLAY

In the following pages can be seen some illustrations of the contributions made by members of the Festival Pattern Group who were invited to join too late for the initial displays, and also some further applications of materials. In addition, we illustrate the Regatta Restaurant at the South Bank Exhibition, the main display for the whole scheme. It is not possible to show in the illustrations all the individual contributions, but a good idea is given of some of the practical results of the scheme. The Group display, which is in the foyer of the Regatta Restaurant (and also in the Land Travelling Exhibition) shows at a glance the complete range and represents an admirable contribution to the long list of well-designed and well-produced products made by British industry today.

ACKNOWLEDGMENT: The colour illustration of Dunlop sheeting (p. 4) is reproduced by courtesy of *The Queen*

HERE THE DECORATIONS COME FROM CRYSTAL STRUCTURES

In this Restaurant a novel system of decorative pattern has been used. The patterns applied to the decorations, furnishings and tableware are derived from crystal structure diagrams, which are the maps that a scientist draws to record the arrangements of the atoms in particular materials under examination.

This branch of science (crystallography) is highly developed in Britain and the suggestion to put the diagrams to decorative uses came originally from the crystallographers themselves. An explanation of the science appears in the Dome of Discovery in "The Physical World" and some of the decorative applications are displayed there too.

A group of leading manufacturers has co-operated in this experiment and this display shows examples from all their work, which extends beyond the decoration of this restaurant, to many other products decorated with these exciting patterns.

Descriptive panel, border design, in colour, based on Polythene. In the group display at the Regatta Restaurant, South Bank. London Typographical Designers Ltd

Chair. In the Land Travelling Exhibition. Made by Goodearl Bros Ltd, designer E. L. Clinch. Ornament based on Polythene diagram

BELOW: *Tea set, R. H. & S. L. Plant Ltd, designer Miss Hazel Thumpston. A general essay on the Crystal Patterns theme. In the group display at the Regatta Restaurant, South Bank*

Menu and Wine List covers for the Regatta Restaurant, South Bank. Motif derived from Hydrargillite diagram. London Typographical Designers Ltd

BELOW: Cinema seats upholstered with PVC coated fabric, a product of Imperial Chemical Industries Ltd, Leathercloth Division (Fabric designer: C. Garnier). In the cinema at the Exhibition of Science, South Kensington

Wall covering in nitrocellulose-coated fabric. Imperial Chemical Industries Ltd, Leathercloth Division (Fabric designer: C. Garnier). Insulin diagram. At the Exhibition of Science, South Kensington

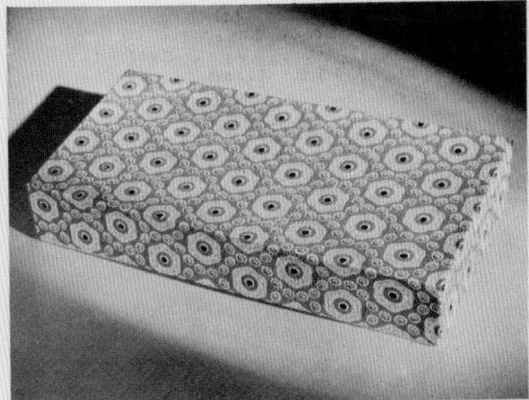

Box paper from Spicers Ltd, designer W. Farquhar, based on Haemoglobin diagram. In the group display at Regatta Restaurant, South Bank

ABOVE: *The interior of the Regatta Restaurant, South Bank. Here can be seen the general effect of some of the materials and designs used in combination*

BELOW: *Settees upholstered in PVC coated fabric. Imperial Chemical Industries Ltd, Leathercloth Division (Fabric designer: C. Garnier). At the Exhibition of Science, South Kensington*

MEMBERS OF
THE FESTIVAL PATTERN GROUP

Barlow & Jones Ltd
E. Brain & Co Ltd
British Celanese Ltd
Carter & Co Ltd
Chance Bros Ltd
Dobroyd Ltd
Dunlop Rubber Co Ltd
Elkington & Co Ltd
General Electric Co Ltd
A. C. Gill Ltd
Goodearl Bros Ltd
C. A. Harvey & Co (London) Ltd
Imperial Chemical Industries Ltd
                    (Leathercloth Division)
Arnold Lever (London) Ltd
John Line & Sons Ltd
London Typographical Designers Ltd
Old Bleach Linen Co Ltd
R.H. & S.L. Plant Ltd
Spicers Ltd
Stevens & Williams Ltd
James Templeton & Co Ltd
Vanners & Fennell Bros Ltd
Vernons Industries Ltd
Warerite Ltd
Warner & Sons Ltd
Josiah Wedgwood & Sons Ltd
Wood Bros Glass Co Ltd

A linoleum manufacturing company
Addresses of members may be had upon application to Mr Hartland Thomas, at the Council of Industrial Design.

16

Another view of the Regatta Restaurant

BELOW: *The group display in the foyer of the Regatta Restaurant, South Bank. In this panel are shown representative examples from each of the members of the Festival Pattern Group. Attention is drawn to the Wedgwood group (designer: Peter Wall), centre right, and to the plate from E. Brain & Co Ltd (designers: Miss H. Thumpston and Peter Cave), centre, not illustrated elsewhere*

CHAPTER THREE

FESTIVAL PATTERN GROUP: CATALOGUE OF MANUFACTURERS AND DESIGNS

AAD – Archive of Art and Design, V&A (Helen Megaw Papers)
COID – Council of Industrial Design
DCA – Design Archives, University of Brighton (Design Council Archive: Festival Pattern Group files)
HM - Helen Megaw
FPG – Festival Pattern Group
MAG – Manchester Art Gallery
MHT- Mark Hartland Thomas
NCM – Nottingham Castle Museum
SM – Science Museum
V&A – Victoria & Albert Museum
VAN – Vanners Archive
WAG – The Whitworth Art Gallery, The University of Manchester
WTA – Warner Textile Archive

BARLOW & JONES

Tradename: Osman
FPG code: A501

Textile manufacturers based in Manchester. Originally founded in Bolton by James Barlow in the 19th century, Barlow & Jones was a large textile company with several mills, specialising in household textiles and dress fabrics. Their Chairman, Sir Thomas Barlow, was the first Chairman of the COID. When initially approached about the FPG, however, he was not particularly supportive. H. Swift attended the first FPG meeting, 16/12/1949. 'Swift of B&J: preparing to make prints for towels, bedspreads & dress, roller printing not screen, some spun rayon, within a month or two something to show.' (MHT file note, 2/1950).

One of Barlow & Jones' textiles was used in the Regatta Restaurant; others were shown in the Dome of Discovery. Some were commercially available in May 1951. 'The furnishing fabric may cost as much as 25/- per yard, 48", in the shops including uplift and purchase tax. / We are putting the marocain into utility so this will only cost about 6/- per yard. Organdie will have to be non-utility and may cost 9/11d per yard in the shops.' (Letter from B&J to HM, 7/5/1951).

Barlow & Jones declined to participate in the FPG after the Festival: 'The suggestion of carrying on the Group offers certain attractions but, after full discussion, we have reluctantly come to the conclusion that we could not contribute much to the Group, nor can we receive much benefit from membership of it ourselves. We have not evidence to show that our co-operation in the Group's work has been of advantage to our business.' (Letter from E. Cartwright to MHT, 18/10/1951).
Archives: AAD 1977/3/205; DCA 5384-1, 5384-4, 5396

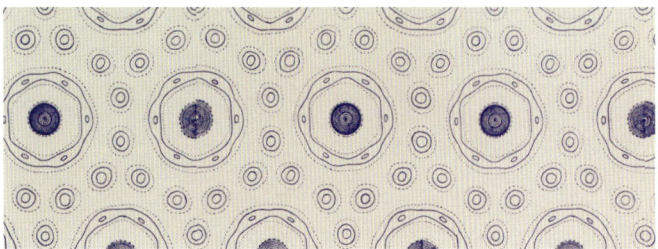

Haemoglobin 8.26 diagram

FPG 1
Haemoglobin 8.26 furnishing fabric
Jacquard-woven cotton and yarn-dyed viscose rayon
Crystallographer: Max Perutz
w.122cm; *repeat*: 6.3 x 10cm; 7 colourways
Collections: V&A Circ.66-1968+A; T.446F-1977
Souvenir Book, Cover and p.2; *British Textiles*, pp.50-51; *Skinner's Record*, p.475
Limited production in 1951. Bronze colourway used in alcove in Regatta Restaurant (100 yards); also shown in FPG display. An enlarged version may have been produced as a bedcover fabric.

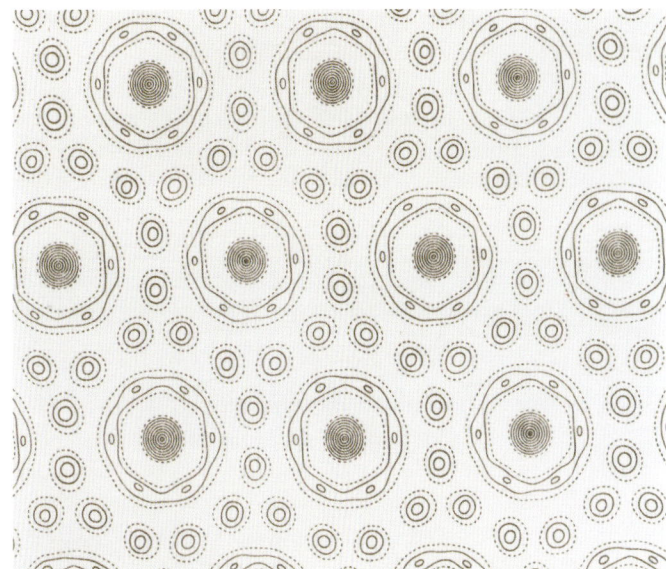

FPG 2
Haemoglobin 8.26 dress fabric
Roller-printed viscose rayon marocain
Crystallographer: Max Perutz
Pattern no: R8429/2156; w.91.5cm
Collections: V&A T.446F-1977
Skinner's Record, p.475
Limited production in 1951 under the Utility scheme.

FPG 3
Haemoglobin 8.26 dress fabric
Roller-printed Egyptian cotton cambric and cotton organdie
Crystallographer: Max Perutz
Pattern no: C9041; w.91.5cm; *repeat:* 5.7 x 9.5cm
Collections: V&A Circ.77-1968+A; T.446F-1977; SM 1976-644/15
Souvenir Book, p.2; *Architectural Review*, p.238
Limited production in 1951. Organdie proposed for waitresses' aprons, but probably not used.

E. Brain & Co.

Tradename: Foley Bone China
FPG code: A529

Ceramics manufacturers based in Fenton, Stoke-on-Trent. Founded in 1885 by E. Brain and his son, W.H. Brain; succeeded in 1931 by the latter's son, Eustace William Brain. Initially specialised in bone china tea and breakfast wares, but after World War II diversified into dinner ware and hotel ware. Merged with Coalport in 1963; taken over by Wedgwood Group in 1967; factory closed in 1992.

E. Brain & Co. were invited to participate in the FPG at a very late stage: 'we are finding this work very amusing and interesting, and should be very pleased to join the FPG.' (Letter from E.W. Brain to MHT, 27/2/1951). Hazel Thumpston and Peter Cave had recently joined the company as trainee designers from the Royal College of Art, where they had designed crystal structure prints for ceramics in a project with Professor R.W. Baker. Their design was not commercially produced due to shortage of time and post-war manufacturing restrictions.
Archives: DCA 5384-3, 5396

Aluminium Hydroxide 8.8 diagram

FPG 4
Aluminium Hydroxide 8.8 plate
Bone china, printed in overglaze enamels, gilded rim and banding
Crystallographer: Helen Megaw
Designers: Hazel Thumpston and Peter Cave
d.26.7cm
Collections: V&A Circ.40-1968
Prototypes only. Shown in FPG displays in Regatta Restaurant and Land Travelling Exhibition.

British Celanese

Tradename: Celanese Fabrics
FPG code: A503

Manufacturers of textiles, plastics and chemicals; headquarters at Celanese House, Hanover Square, London. The British Cellulose & Chemical Manufacturing Co. was established in 1916 by the chemist Henri Dreyfus as an offshoot of a Basel-based company Cellonit Gesellschaft Dreyfus & Co, pioneers of cellulose acetate. In 1918 the name British Celanese was adopted. Commercial production of acetate yarns began in 1921.

British Celanese were proposed for the FPG by Dennis Lennon, Director of the Rayon Industry Design Centre, 29/7/1949. W.A. Dickie, Director, 'was worried, being a chemist, about the association of ideas between the origin of the designs and the finished product. For example, he said, "Do you think any girl will want to wear a diagram of insulin, for instance?"' (Letter from Dennis Lennon to MHT, 29/7/1949).

He and W.H. Tomlinson, Woven Fabrics Manager, attended the first FPG meeting, 16/12/1949. Tomlinson advised Misha Black that they would need 6 months' run-in time to produce fabrics for the restaurant. None were selected in the end, but two were shown in FPG displays in Regatta Restaurant and Dome of Discovery.

Archives: AAD 1977/3/180; DCA 5384-1, 5384-3, 5384-4, 5396

Afwillite 8.45 diagram

FPG 5
Afwillite 8.45 dress fabric
Screen-printed spun rayon
Crystallographer: Helen Megaw
Designer: S.M. Slade
Pattern no: 449/6003; *w*.91.5cm; *repeat*: 30.5 x 16.5cm; 4 colourways
Collections: V&A Circ.75-1968 +ABC; SM 1976-644/25
Souvenir Book, p.6: 'The Celanese examples... make play with the contour type diagram which is like the free-shape outline of current fashion'; *British Textiles*, p.55; *Queen*; *Skinner's Record*, p.475
Limited production in 1951. 37s 6d per yard. Shown in FPG displays in Regatta Restaurant and Land Travelling Exhibition. HM was given a 5yd length.

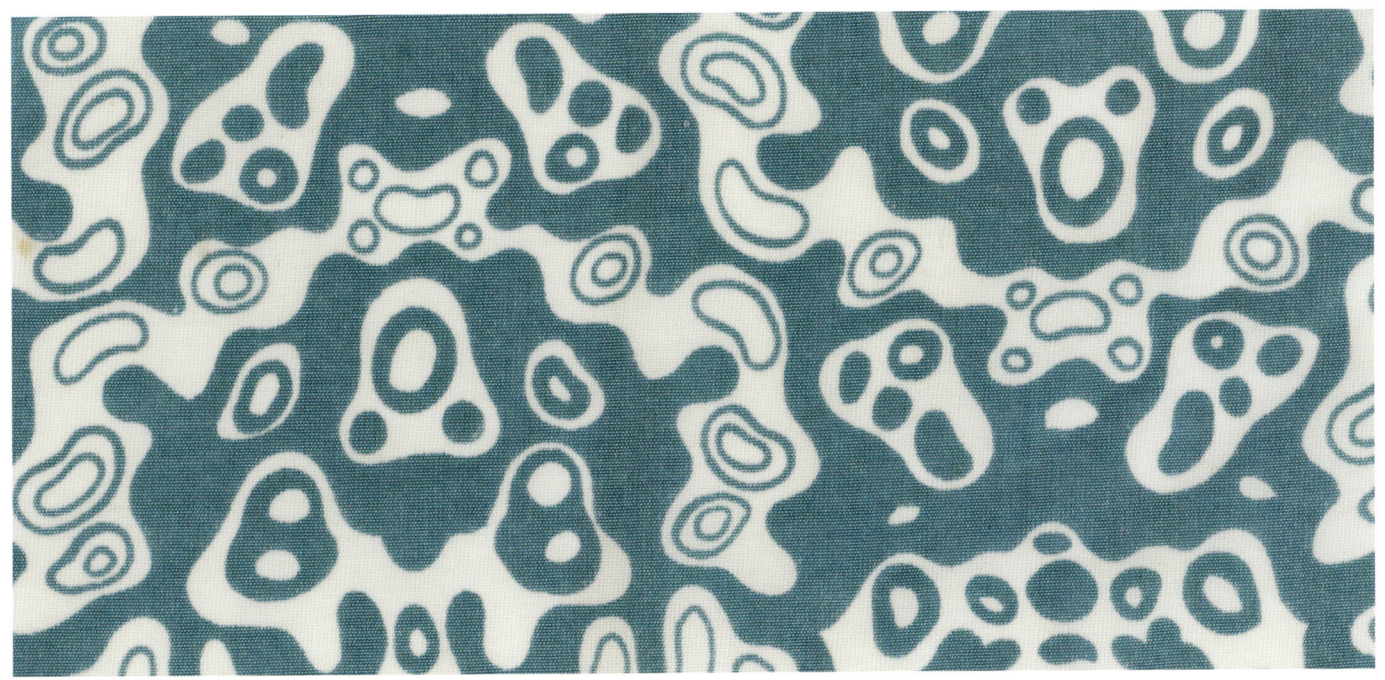

Horse Methaemoglobin 8.23 diagram

FPG 6
Horse Methaemoglobin 8.23 dress fabric
Printed fine filament acetate rayon crepe
Crystallographer: Max Perutz
Designer: S.M. Slade
Pattern no: 516/6004; *w.*91.5cm; *repeat*: 13.5 x 9.2cm; 5 colourways
Collections: SM 1976/644/31
Souvenir Book, p.7; *British Textiles*, p.55; *Skinner's Record,* p.475
Limited production in 1951. 25s per yard. Shown in FPG display in
Land Travelling Exhibition. Gisela Perutz, wife of Max Perutz, had a
dress made from this fabric.

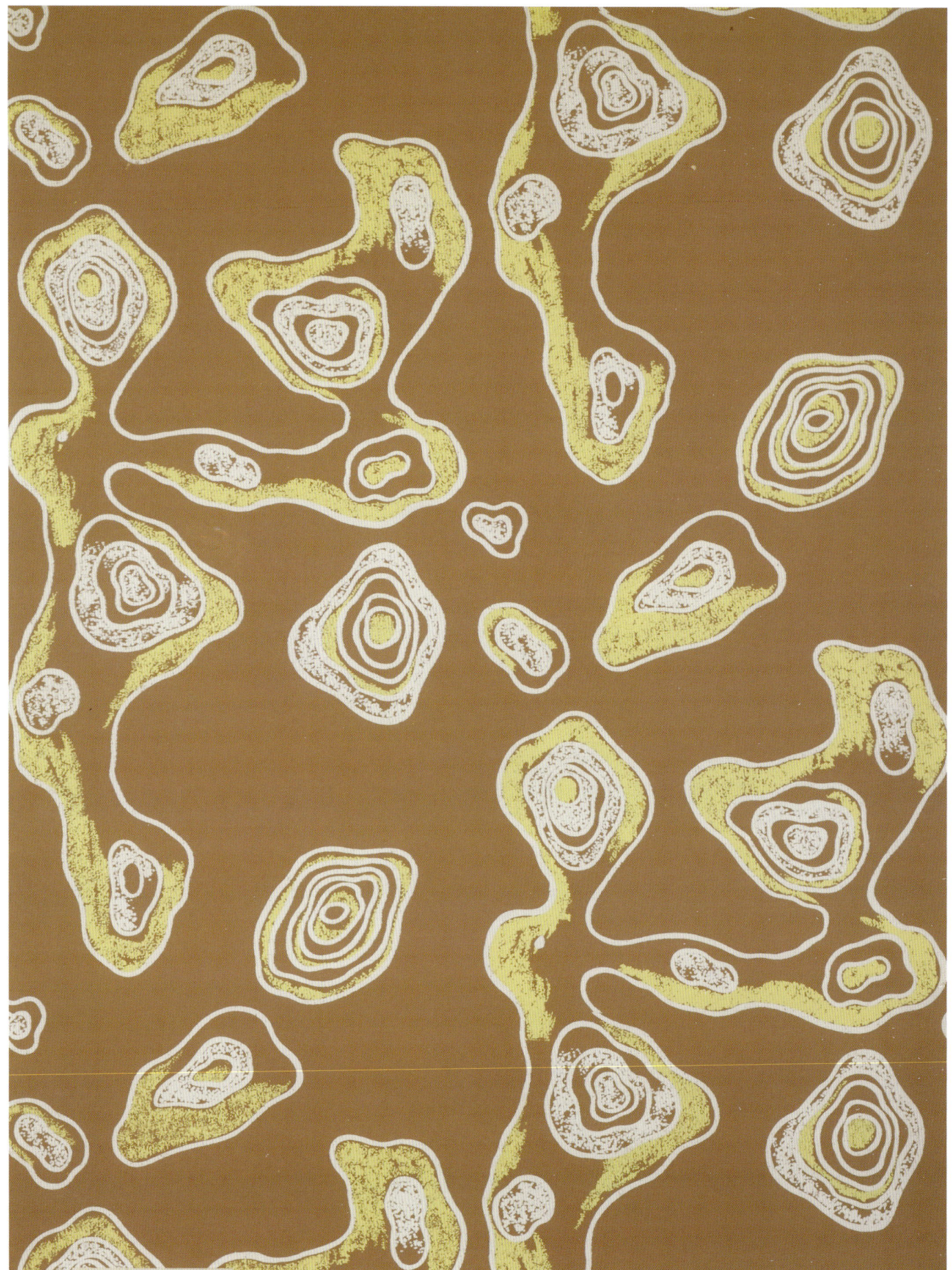

FPG 35 - Afwillite 8.45 wallpaper, designed by William Odell for John Line and Sons

CARTER & CO.

Tradename: Poole Pottery
FPG code: A517

Ceramics manufacturers based in Poole, Dorset. Founded in 1873, Carter & Co. was the tile-making parent company of Carter, Stabler & Adams, established in 1921, later known as Poole Pottery. Taken over by Pilkington's Tiles in 1964; Poole Pottery re-established its independence in 1992 following a management buy-out, but the factory closed in 2006.

Lucien Myers, Managing Director of CSA was approached about the FPG, 29/9/1949. Cyril Carter responded on behalf of Carter & Co., 1/11/1949. David Carter, a London-based architect who had recently overseen the redesign of the factory, attended the first FPG meeting, 16/12/1949. Two drawings were presented, 11/7/1950. Reginald Till, head designer, was responsible for the final tile panel, featuring a small section of a crystal structure diagram, greatly enlarged.

David Carter expressed a strong desire to remain involved after the Festival: 'Because our output of decorated tiling is necessarily quite small and most of it is required for work specifically designed for individual buildings, we were not then in a position to produce standard patterns using crystalline designs. We could only show our keenness to execute work in this idiom and our ability to do so. However, we have been able to expand our decorating department during the last six months, and it will be of great value to us to learn the public reaction to crystalline motifs.' (Letter from David Carter to MHT, 2/10/1951).
Archives: DCA 5384-3, 5384-4, 5396

FPG 7
Zinc Hydroxide 8.39 tile panel
24 hand-painted earthenware tiles
Crystallographer: Helen Megaw
Designer: Reginald Till
Collections: V&A Circ.1968-38
w.61 x h.91.5cm
Souvenir Book, p.11
Produced as a limited edition of three, intended for FPG displays in Regatta Restaurant, Exhibition of Science and Land Travelling Exhibition.

CHANCE BROTHERS

Tradename: Chance Glass
FPG code: A516

Glass manufacturers based in Smethwick, Birmingham. Founded in 1832 when glassmaker Robert Lucas Chance went into business with his brother William. Initially focused on window glass and optical glass; later expanded into scientific glass, illuminating glass, heat-resistant ovenware and pressed glass tableware. Taken over by Pilkington Brothers in 1945, but retained their own name and identity. Factory closed in 1981.

J.W. Chance, London Manager, was approached about the FPG on 18/8/1949 and attended the first FPG meeting, 16/12/1949. Initial designs by their staff designer, Mr Harwood, were considered unsatisfactory, so in May 1950, at the instigation of MHT, an independent industrial designer, J. Beresford Evans, was employed instead. 'Chance Bros., one of the members has so far not been able to get a satisfactory design for figured rolled glass and has asked me to suggest a designer. I hope you will have a shot at it... / There is some urgency in the matter as the hand engraved rollers take nine months to make.... / The repeat must be such that it is economical in cutting... / The solution that I have suggested... is to use the diagrams at a minute scale... then it does not matter how you line up the pattern or, if it does matter, you never cut very much to waste.' (Letter from MHT to J. Beresford Evans, 12/5/1950).

Beresford Evans responded enthusiastically: 'I had been excited by an electron density map that Bragg showed at the RSA in 1937, but got no further than making a landscape of it...

/ Do you think it would be possible for me to meet Dr Megaw? For "fine art" purposes I want more information on electron density than I have yet been able to collect.' (Letter from J. Beresford Evans to MHT, 15/5/1950). Evans submitted three designs, including Cristobalite and China Clay, but Apophyllite was chosen and subsequently mass-produced. Chance were so pleased from a technical and commercial point of view that they later commissioned Evans to design three further glass patterns. HM also approved: 'I am delighted to hear that our effort has followed so faithfully the original diagram and that, even more important, the result meets with your aesthetic approval.' (Letter from J.W. Chance to HM, 29/11/1950).
Archives: AAD 1977/3/175; DCA 5384-1, 5384-2, 5384-3, DCA 5384-4, 5396

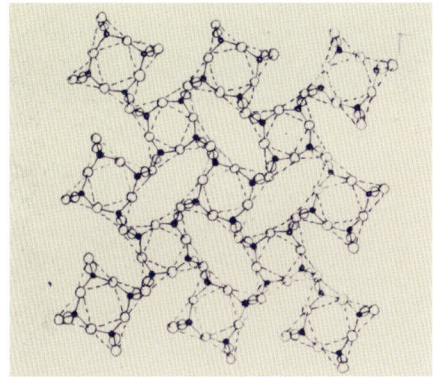

Apophyllite 8.30 diagram

FPG 8
Apophyllite 8.30 window glass ('Festival')
Figured rolled sheet glass
Crystallographer: W.H. Taylor
Designer: J. Beresford Evans
Repeat: 1.2 x 1.2cm
Collections: AAD 2000/4/55
Souvenir Book, p.10: 'This design is as unobtrusive as the random-pattern obscured glasses used most frequently today; but it has a delicate charm as well, which comes from the subtlety of the exquisite little design. Two pitfalls in window-glass have been avoided – to prevent cutting to waste for matching, the pattern is on a slant and the repeat small; and so that sunshine does not burn the curtains, the blobs themselves are small.'

Mass-produced under the name 'Festival' from 1951 onwards. Shown in FPG displays in Regatta Restaurant, Dome of Discovery and Land Travelling Exhibition. Used for screens in women's lavatories at Regatta Restaurant. At the Exhibition of Science it was used for shelves in the buffet amd for screens in the reception suite and entrance to the cinema. Also used in Homes and Gardens Pavilion and Battersea Beer Garden. HM installed 'Festival' glass in two kitchen doors in her house in Cambridge. Panels were also installed in the toilets in the Chemistry Department at Cambridge.

DOBROYD
FPG code: A503

Textile manufacturers, specialising in woollen dress fabrics, based in Huddersfield, West Yorkshire. William Haigh, Director, was approached about the FPG on 2/2/1950. Designer Tony Dawson attended FPG meetings from 16/3/1950. Translating crystal structures into woven woollen cloth proved challenging, but some interesting effects were created using fancy weaves and novelty yarns. Dobroyd's cautious approach limited their commercial impact, however, as they were reluctant to put the fabrics into full-scale production (minimum order 60-70 yards). After the Festival they said they no longer wished to be involved: 'we already have a staff of designers and are continually bringing out our own productions. We are different to many other businesses in that respect, but it is the general practice throughout the woollen and worsted trade to have a highly developed and skilled staff of designers.' (Letter from William Haigh to MHT, 2/10/1951).
Archives: AAD 1977/3/209; DCA 5384-2, 5384-4, 5396

FPG 9
Mica 8.35 [2] dress fabric
Two-colour woven wool
Crystallographers: W.W. Jackson / J. West
Designer: Tony Dawson
Pattern nos: D112/199; SB 4Y/1825; repeat: 1.3 x 5cm; 3 colourways
Collections: V&A Circ.68-1968+ABC; SM 1976-644/39
Souvenir Book, p.8; *British Textiles*, p.51
Samples only. Shown in FPG display in Regatta Restaurant.

FPG 10
Mica 8.16 dress fabric
Lightweight woven wool with raised surface weave
Crystallographers: W.W. Jackson / J. West
Designer: Tony Dawson
Pattern no: SB4Y/1805; repeat: 1.3 x 2.2cm; 1 colour
Collections: V&A Circ.69-1968; SM 1976-644/37
Souvenir Book, p.8; *British Textiles*, p.51
Samples only. Although closest to Mica 8.16, this design also resembles Mica 8.35 [3].

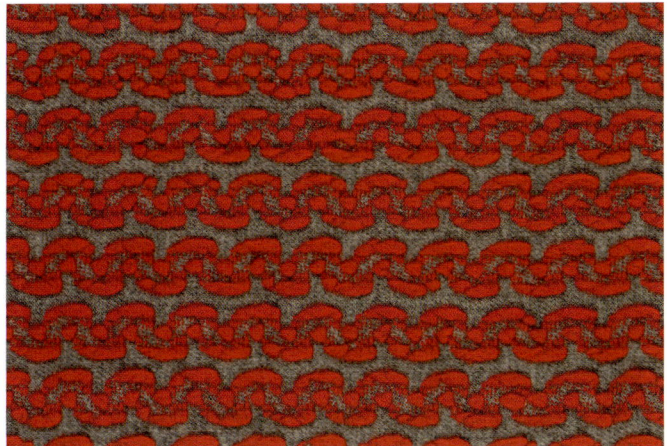

FPG 11
Polythene 8.59c dress fabric
Two-colour woven wool with novelty weave
Crystallographer: Charles William Bunn
Designer: Mildred Taylor (attributed)
Pattern no: F1/219; w.147cm; 1 colourway
Collections: V&A Circ.67-1968
Samples only. Not itemised in FPG documentation, but included in COID's donation to V&A.

FPG 12
Polythene 8.59c dress fabric
Woven merino wool with novelty weave
Crystallographer: Charles William Bunn
Designer: Mildred Taylor
Pattern no: SB4Y/1807; 1 colour
Collections: V&A Circ.74-1968; T.446F-1977; SM 1976-644/38
Applications: Samples only.

DUNLOP RUBBER COMPANY
FPG code: A509

Rubber manufacturers based at Fort Dunlop, Erdlington, Birmingham. Founded by John Boyd Dunlop in 1888 to manufacture his newly invented pneumatic rubber bicycle tyres, Dunlop soon became a major producer of car tyres as well. By the mid 20th century the company had diversified into many other products, ranging from Wellington boots, rainwear and hot water bottles, to conveyor belts, vinyl tiles and Dunlopillo mattresses.

Dunlop were approached about the FPG on 18/8/1949, initially with a view to creating rubber flooring for the restaurant, eventually used at the Exhibition of Science. Robert Cantor from the Industrial Design Section attended the first FPG meeting, 16/12/1949. Dunlop's main focus was a new experimental product, printed PVC sheeting. Prototypes were shown in FPG displays in Regatta Restaurant and Dome of Discovery using 'pilot plant production', but the products themselves were not commercially available, 'because the method of colour printing employed is only just emerging from the development stage.' (Letter from E.A. Murphy, Development Section, to HM, 17/5/1951). A few months later these problems were still unresolved: 'The type of printed p.v.c. sheeting exhibited in the Festival Pattern Group was very much a prototype illustrating the sort of material we are hoping to produce within the next six to nine months. The samples were only made on a very small laboratory plant.' (Letter from N.G. Bassett Smith, Manager, Flexible Plastics, Dunlop Special Products, to Sherman M. Fairchild, 30/7/1951).
Archives: AAD 1977/3/166, 211; DCA 5384-3, 5384-4, 5396

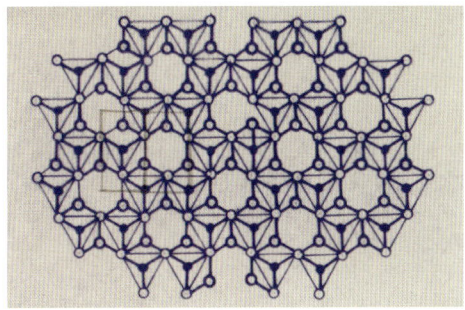

China Clay 8.6 diagram

FPG 13
China Clay 8.6 PVC sheeting
Printed polyvinyl chloride
Crystallographer: G.W. Brindley
Designer: Mary A. Harper
Collections: V&A Circ.58-1968
Prototypes only.

FPG 14
Insulin 8.27 PVC sheeting
Printed polyvinyl chloride
Crystallographer: Dorothy Hodgkin
Designer: Mary A. Harper
Repeat: 24 x 27.3cm
Collections: V&A Circ.57-1968
Souvenir Book, p.4; *Queen*
Prototypes only. Shown in FPG display in Regatta Restaurant.

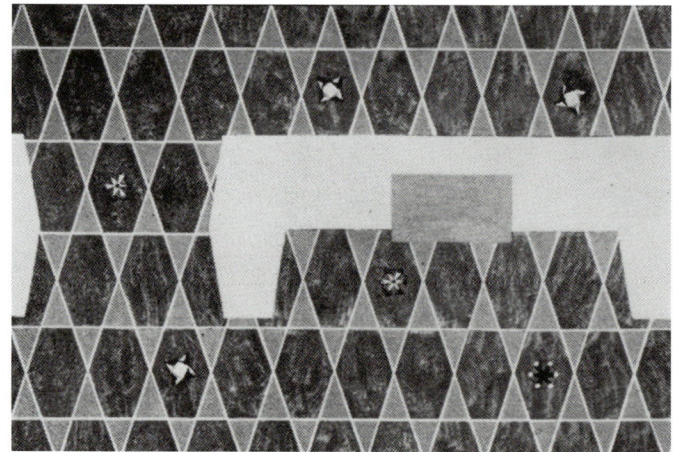

FPG 15
Mica 8.35 [2] vinyl flooring
Multicoloured vinyl floor tiles
Crystallographers: W.W. Jackson / J. West
Architectural Review, p.239
Produced as a one-off for the buffet at the Exhibition of Science. The commission is mentioned in a letter from Robert Cantor to MHT, 17/11/1950. The suppliers, Semtex Ltd., were part of Dunlop.

ELKINGTON & CO.

FPG code: A524

Manufacturers of cutlery and electroplate based in Birmingham. Founded by George Richards Elkington and Henry Elkington around 1830, Elkington & Co. took out a patent for electroplating in 1840 which established their worldwide fame. By the early 20th century they were producing large quantities of cutlery for ocean liners. They were invited to join the FPG at a late stage after the withdrawal of Walker & Hall. Hagenbachs asked them to produce cutlery, cake stands and bread baskets for the Regatta Restaurant in January 1951, but they declined due to shortage of time. Prototype cutlery was featured in FPG displays, but could not be manufactured because of materials shortages.

'The decoration is taken from one of your prints of Polythene, and while I am not very satisfied with it, it was probably the best of the four or five attempts which we made... / We have not offered it for sale, as we are not quite sure whether we intend to produce it at the present time...You probably know that three of the metals which we use and which form the basis of all our articles, namely copper, zinc and nickel, are in such short supply that at the end of June we are not permitted to manufacture any more articles of Holloware, such as Tea Sets, etc., and while flatware (spoons, forks, knives) is not prohibited, there is the greatest difficulty to-day to obtain materials. However, when things improve, we shall probably add this to our range of patterns.' (Letter from Elkington to HM, 9/5/1951).
Archives: AAD 1977/3/213; DCA 5384-3, 5384-4, 5396

FPG 16
Polythene 8.59d cutlery
Electro-plated nickel silver with stamped decoration
Crystallographer: Charles William Bunn
Designer: H.G. Bowring
Souvenir Book, p.7: 'It is easier to spoil silverware by decoration than to improve it. Here the designer has graduated the pattern in sympathy with the profile and made a decorated handle that is as good as, if not better than, the plain one.'
Prototypes only. Shown in FPG displays in Regatta Restaurant and Land Travelling Exhibition.

FPG 71 - Afwillite 8.44 furnishing fabric ('Surrey'), designed by Marianne Straub for Warner & Sons

GEC

FPG code: A504

Manufacturers of electrical products and lighting; headquarters at Magnet House, Kingsway, London. Founded in 1886 as the General Electric Apparatus Company, GEC initially focused on the manufacture of telephones, electric bells and lighting. They played a significant role during World War I in developing radios and signalling lamps, and again during World War II through supplying electrical and engineering products to the military. During the 1920s GEC were involved in the creation of the national grid.

GEC were approached about the FPG on 27/2/1950 with a view to producing a lighting feature for the Regatta Restaurant. Their managing director, L. Gamage, was chairman of the COID's Industrial Committee. Hartland Thomas proposed that the DRU 'might design (on a crystal structure theme) a modern equivalent of the crystal chandelier with all the curious effects of fluorescent tubes, light piped through Perspex and motorise the whole thing as a mobile!' (Letter from MHT to Misha Black, 22/2/1950). HM approved: 'I like the idea of these structures being turned into exciting lighting fittings.' Their designs were rejected by the DRU, however, and they ended up producing wall brackets instead. GEC also produced a dramatic lighting installation for the Exhibition of Science inspired by ball-and-spoke atomic structure models, designed by Brian Peake, although this was not part of the FPG scheme. (see fig.3)

Archives: DCA 5384-1, 5384-3, 5384-4, 5396

FPG 17
Beryl 8.9 double wall bracket lamp
Plastic laminate back plate, lace patterned lampshade, brass brackets
Crystallographer: Lawrence Bragg
Designer: R.J. Reynolds
Souvenir Book, p.1: 'These pictures epitomise the spirit of co-operation in the Group; the pattern in the shade is the actual lace shown alongside it... and the same pattern is echoed in the plastic backplate by a third member, Warerite Ltd.'; *Architectural Review*, p.238; *Homes and Gardens*, p.34; *Queen*

Prototypes only, incorporating materials supplied by two other FPG members, Warerite and A.C. Gill. Shown in FPG display in Land Travelling Exhibition. Used in Section SB17 at the Festival (exhibit no.1721), in spite of the architect's distaste. 'I think we could get over the difficulty of the lace pattern on the Warerite mounting by having it a fairly pale colour, pale grey or white for instance. As the fitting will be quite a distance from the spectators I do not think the design on it will then be very apparent.' (Letter from Patience Clifford to MHT, 8/1/1951).

A.C. GILL

Tradename: Witchcraft Laces
FPG code: A523

Lace manufacturers based in Nottingham, specialising in dress and curtain fabrics, trimmings and veils. Originally known as Gill & Bass, the company was founded in 1888 by Albert Charles Gill and William Bass to produce silk veils and Spanish lace. Renamed A.C. Gill in the 1890s, it became a limited company in 1915, with J.G. McMeeking as co-director, joined by his brother A.C. McMeeking in 1920. Production expanded over the next two decades to include dress flouncings, lingerie laces and bridal veils; rayon was introduced and a subsidiary established producing hosiery and hairnet. During the war production switched to parachutes, flags and mosquito nets. The company expanded rapidly from the mid 1940s, taking over several other local firms; later they introduced the world's first all-Nylon lace. During the 1950s the main customers for 'Witchcraft Laces' were makers of bridal gowns. Distribution was handled by their associate firm, Munro Laces. Over time the company diversified further, eventually moving away from lace altogether. In 1980 they acquired the Lion Hair Care brand. A.C. Gill still exists today as the parent company of the Nottingham Textiles Group, created in 2000, whose products range from hairnets to face masks to fly nets.

J.G. McMeeking responded enthusiastically after being approached about the FPG in September 1950 and attended FPG meetings from 16/11/1950. 'Fortunately on the embroidery machines we can re-produce the designs quite faithfully. Where additional lines have been introduced, we are compelled to do this in order to allow continuous working of the thread.' (Letter from J.G, McMeeking to MHT, 23/10/1950). A wide range of crystal structure designs was shown at the Festival, mainly by the company's in-house designer, H. Webster. They entered production and were subsequently shown at the British Industries Fair. The company remained positive to the end: 'The actual sales to the public have been small but undoubtedly participation in your Group has added prestige to this firm and proved first rate publicity material. / The ideas have, of course, provided our designers with a new starting point and we may still see developments resulting from our connection with the FPG.' (Letter from J.G. McMeeking to MHT, 12/10/1951). Some FPG laces were still available as late as 1959, by which date two new crystal structure designs had been added, Oxalic Acid Dihydrate and Myoglobin.

Archives: AAD 1977/3/172-173, 178, 327, 351-369, 577; DCA 5384-2, 5384-3, 5384-4, 5396

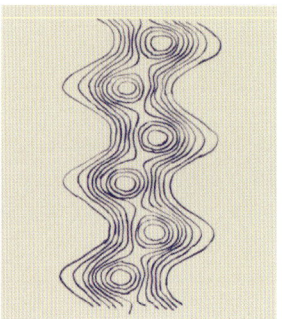

Polythene 8.59c diagram

FPG 18
Apophyllite 8.30 lace
Machine-embroidered cotton lace
Crystallographer: W.H. Taylor
Designer: H. Webster
Pattern no: 5265; w.86.5 - 91.5cm; *repeat*: 7 x 8.2cm
Collections: V&A T.446F-1977; SM 1976-644/23
Manufactured from 1951 onwards. HM was given 3 yards by the manufacturer, 17/10/1951.

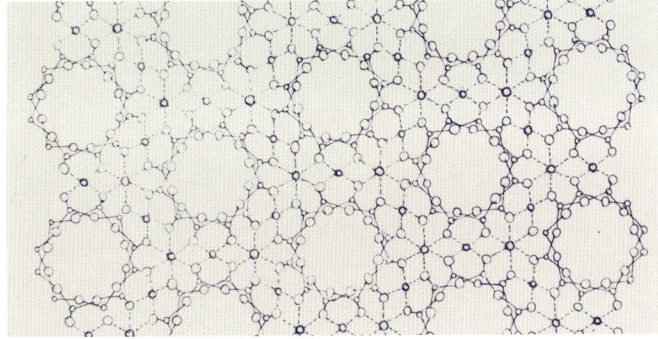

Beryl 8.9 diagram

FPG 19
Beryl 8.9 lace
Machine-embroidered cotton lace
Crystallographer: Lawrence Bragg
Designer: H. Webster
Pattern no: 5197; w.86.5 - 91.5cm; *repeat*: 3.8 x 9.5cm; 2 colours
Collections: V&A Circ.62-1968+A, T.446F-1977; SM 1976-644/19; NCM 1962-43 (dress)
Souvenir Book, p.1; *Queen*; *Skinner's Record*, p.475
Manufactured from 1951; still available August 1959. Used for curtains on east elevation of Regatta Restaurant (280 yards); also shown in FPG display. Incorporated in lampshades on GEC light fittings. Lady Alice Bragg, wife of Sir Lawrence Bragg, wore an evening dress made of Beryl lace at the International Congress of Crystallography in Stockholm, 27 June – 3 July 1951.

FPG 20
Haemoglobin 8.26 lace
Machine-embroidered cotton lace
Crystallographer: Max Perutz
Designer: H. Webster
Pattern no: 5189; w. 86.5cm; *repeat*: 7 x 8.2cm; 3 colours
Collections: V&A Circ.64-1968; V&A Circ.66-1968+A; T.446F-1977; SM 1976-644/16 + 24
Souvenir Book, p.3; *British Textiles*, p.54; *Skinner's Record*, p.475
Manufactured from 1951 onwards; still available in August 1959.

FPG 21
Haemoglobin 8.26 lace
Machine-woven lace (15% silk, 85% delustra rayon)
Crystallographer: Max Perutz
Designer: Arnold Parker of Messrs E. Syon & Son
Pattern no: 5263; w. 86.5cm; 1 colour
Collections: V&A Circ.63-1968; T.446F-1977; SM 1976-644/22
Applications: Manufactured from 1951 onwards; still available in August 1959.

FPG 22
Polythene 8.59c lace
Machine-embroidered cotton lace
Crystallographer: Charles William Bunn
Designer: H. Webster
Pattern no: 5190; *repeat*: 2.5 x 3.8cm
Collections: V&A T.446F-1977; SM 1976-644/17
Souvenir Book, p.7; *British Textiles*, p.54; *Skinner's Record*, p.475
Manufactured from 1951 onwards; still available in August 1959.

FPG 23
Insulin 8.27 lace
Machine-embroidered lace veil
Crystallographer: Dorothy Hodgkin
Designer: H. Webster
Souvenir Book, p.4: 'A free adaptation using part of the diagram as an isolated motif.' *Skinner's Record*, p.475
Intended for bridal veils; possibly manufactured in 1951.

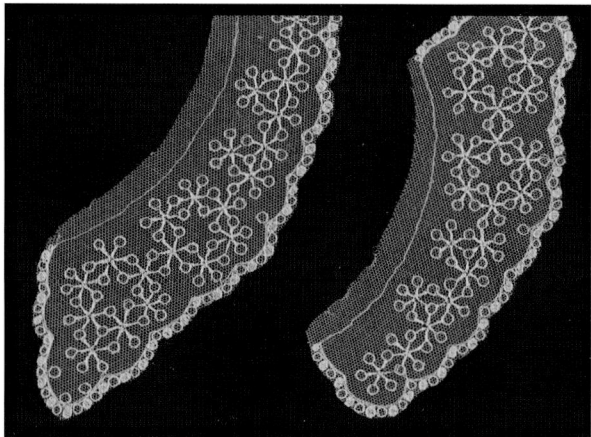

FPG 24
Hydrargillite 8.33 lace
Machine-embroidered cotton lace collars and trimmings
Crystallographer: Helen Megaw
Designer: H. Webster
Pattern no: 5279
Collections: V&A Circ.65-1968, T.446F-1977; SM 1976-644/21
Manufactured from 1951 onwards. 'I think you will be interested to know that the collar with the narrow lace attached is being used in the Regatta Restaurant by the waitresses and they look most attractive.' (Letter from J.G. McMeeking to HM, 17/5/1951). Trimmings used on special menus, including the Festival Dickens Dinner on 24 July 1951.

GOODEARL BROTHERS
FPG code: A528

Furniture manufacturers based in High Wycombe, Buckinghamshire. Grew out of a chair-making business established by four brothers, Henry, Richard, Benjamin and Arthur Goodearl around 1887, although the company was not officially incorporated until 1907. Windsor chairs and cane seat chairs were the main focus initially; later three-piece suites as well. By the 1980s Goodearl had diversified into kitchens and bedrooms; in 1999 they became Whiteleaf Furniture Ltd. The factory closed in 2001, but following a management buyout the company was reborn as Whiteleaf Ltd., based at Thame in Oxfordshire; they remain one of the country's leading cabinetmakers.

Goodearl Brothers joined the FPG in January 1950 following the late withdrawal of Ercol; they were chosen for convenience as they were already supplying the furniture for the Regatta Restaurant. MHT proposed to E.L. Clinch, Director (Design & Development) that they might use Apophyllite for canework patterns, but in the end they produced a chair with a tooled leather seat back decorated with the crystal structure of Polythene.
Archives: DCA 5384-3, 5384-4, 5396

FPG 25
Polythene 8.59b chair
Bent wood with tooled leather seat back
Crystallographer: Charles William Bunn
Designer: E.L. Clinch
Souvenir Book, p.14
Prototype only. Shown in Land Travelling Exhibition and possibly used as seating at Information Points. 'The chair in leather with the tooled motif created great attention but unfortunately was too expensive for general production.' (Letter from E.L. Clinch to MHT, 6/10/ 1951).

FPG 7 - Zinc hydroxide 8.39 tile panel, designed by Reginald Till for Carter & Co

G.A. Harvey & Co. Ltd.

FPG code: A519

Metalwork manufacturers based at the Greenwich Metal Works, London. G.A. Harvey was a large industrial metalworking company which undertook a wide range of engineering, manufacturing and finishing work, including perforation and embossing of sheet metal, as well as producing steel furniture.

They were approached about the FPG on 17/4/1950 with a view to producing perforated grilles and waste paper baskets for the Regatta Restaurant. H.E. Cooper, Director, and Mr. Tarvid (possibly a draughtsman) attended FPG meetings from 13/7/1950. In the end no finished products were created, only samples, because the DRU decided that no metal grilles were required.

Archives: AAD 1977/3/215; DCA 5384-2, 5384-4, 5396

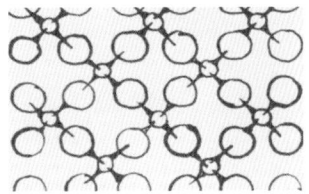

Cristobalite 8.53 diagram

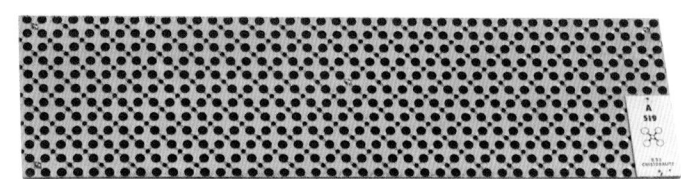

FPG 26
Cristobalite 8.53 pierced metal sheet
Perforated and embossed aluminium
Crystal structure: 'Common knowledge'
Repeat: 2.5 x 2.5cm
Collections: V&A Circ.47-1968; SM 1976-644/14
Souvenir Book, p.8: 'The piercing, which is required by function, necessitates some sort of pattern. In the past, such patterns have been too ornate; nowadays they are perhaps too modest. In the Group's experimental work, Harvey's did several that were more ambitious than this. Public interest in the Festival Pattern Group may encourage them to make tools for some of them.'
Samples only. Shown in FPG displays in Regatta Restaurant and Land Travelling Exhibition.

ICI Leathercloth Divison

Tradenames: Mural Rexines and Vynides
FPG code: A506

Manufacturers of chemicals, specialising in dyestuffs, alkalis, metals, explosives, fertilisers, paints and plastics; headquarters at Nobel House, Buckingham Gate, London. Imperial Chemical Industries Ltd. was formed in 1926 following the merger of four major British chemical companies: Nobel Industries Ltd.; Brunner, Mond & Co.; United Alkali Company; and British Dyestuffs Corporation. Rubberised 'leathercloth' fabrics were introduced shortly afterwards, leading to the formation of a Leathercloth Division at Hyde in Cheshire. By the 1930s ICI had 25 plants and was a key supplier to the military. As well as being a manufacturer, it was also a research organisation, employing many scientists, including crystallographers. ICI's inventions included Perspex and Terylene. Over the years the company has undergone many transformations, moving into fibre-spinning and PVC production in the 1960s, and expanding in pharmaceuticals in the 1980s. In 1993 its bioscience activities (including pharmaceuticals and agricultural chemicals) were split off into a new company, the Zeneca Group. In recent years ICI's chief areas have been paints (notably Dulux brand), adhesives, coatings, polymers, electronic materials and starches. In August 2007 ICI accepted a takeover bid from the Dutch company Akzo Nobel.

W.J. Worboys, Director of ICI, was approached about the FPG on 1/12/1949 regarding their nitrocellulose-coated wallcoverings and upholstery fabrics (Mural Rexines and Vynides), a new product range intended for heavy-duty use that ICI were keen to promote in schools, restaurants and cinemas. The novel feature for the FPG was the application of printed patterns. C.W. Franks, Chairman, and Kenneth Robertson, Commercial Director of ICI's Leathercloth Division both attended the first FPG meeting, 16/12/1949. Charles C. Garnier, an RCA graduate, was engaged to design their patterns. He

subsequently became Division Designer, much to the approval of the COID: 'Mr Garnier has just called upon me and showed me a portfolio of his new designs for you. I think they are very good indeed; I do congratulate you and him. I am putting him in touch with several designers and architects for the Festival and I should like you to give him permission to prepare for me a complete set of samples to show to some of my colleagues here with a view to using the material in various ways.' (Letter from MHT to Kenneth Robertson, 27/9/1950).

Rexines and Vynides were used extensively at the Exhibition of Science. Samples were shown in FPG displays in the Dome of Discovery, Regatta Restaurant and Land Travelling Exhibition. ICI intended to install 4-colour roller-printing equipment for mass-production, but it did not arrive in time, so the FPG samples were stencil-printed. 'Unfortunately, we have only been able to produce these in very limited quantities by improvisation of plant, and until we get the necessary equipment which will enable us to produce the designs in quantity I am afraid that we have nothing available for sale.' (Letter from Kenneth Robertson to HM, 8/5/1951). There may have been further technical hitches, as there is no record of the range ever being manufactured.

In spite of these setbacks, ICI remained positive about the creative benefits: 'When we originally accepted your invitation to participate in the Festival Pattern Group we did so because we were most anxious to co-operate with you and did not really expect that we would profit particularly from this project itself..../ There is no doubt that our participation in the Pattern Group... has served the purpose of stimulating to a remarkable degree our interest in design; the mere fact that we were brought into contact with you in this way served as a sort of catalyst.' (Letter from Kenneth Robertson to MHT, 2/10/1951.

Archives: AAD 3/217-1977; DCA 5384-1, 5384-2, 5384-3, 5384-4, 5396

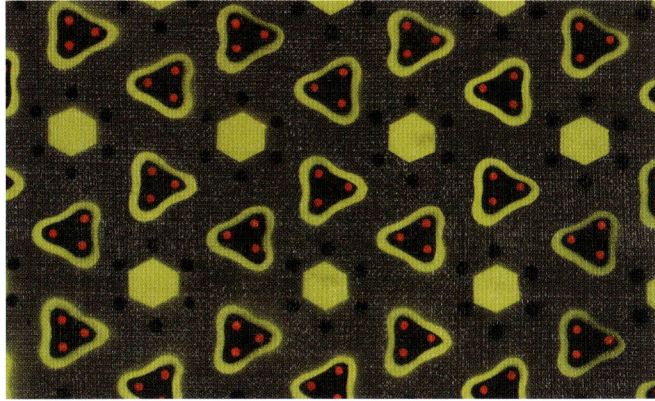

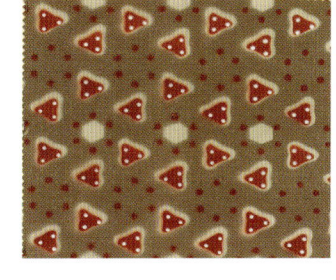

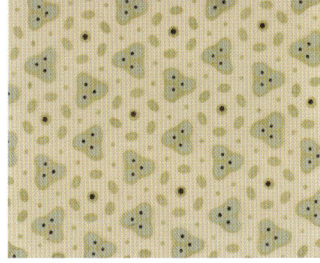

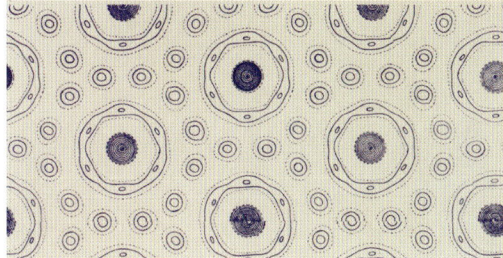

Haemoglobin 8.26 diagram

FPG 27
Haemoglobin 8.26 Mural Rexine upholstery fabric
Nitrocellulose-coated fabric, stencil-printed
Crystallographer: Max Perutz
Designer: Charles C. Garnier
Pattern no: CR8; *repeats*: 31.7 x 22.9cm, 13.3 x 7.6cm, 3.2 x 5.4cm (3 different scales); 9 colourways.
Collections: V&A Circ.52-1968, Circ.54-1968+AB
Souvenir Book, p.3; *Architectural Review*, p.238; *Queen*
Limited production in 1951. CR8/2 was shown in Land Travelling Exhibition; CR8/3 in Dome of Discovery.

FPG 28
Insulin 8.25 Mural Rexine wallcovering and Vynide upholstery
Nitrocellulose-coated fabric, stencil-printed
Crystallographer: Dorothy Hodgkin
Designer: Charles C. Garnier
Pattern nos: CR5 and CR7; *repeat*: 5 x 2.8cm (2 different scales); 12 colourways.
Collections: V&A Circ.52-1968, Circ.53-1968, Circ.55-1968+A
Souvenir Book, p.5 and p.15
Limited production in 1951. Used extensively in Exhibition of Science, including on display stands; CR7/2 used on walls in buffet; CR7/7 used on seats in cinema and ambulance room. CR5/2 and CR5/3 shown in FPG display in Regatta Restaurant. CR7/6 was used in Homes and Garden pavilion (SB/18 A: Child in the Home – Room for a Toddler). In Land Travelling Exhibition CR7/1 used for Enquiry Office; CR7/2 shown in FPG display.

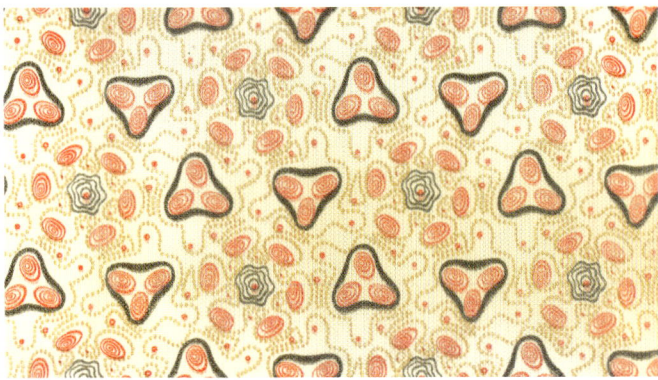

FPG 29
Insulin 8.24 Mural Rexine wallcovering
Nitrocellulose-coated fabric, printed
Crystallographer: Dorothy Hodgkin
Designer: Charles C. Garnier
1 colourway
Collections: V&A Circ.52-1968
Sample in ICI Mural Rexines and Vynides book in V&A; printed on thinner material in a different style. No references to this design in archives.

FPG 31
Myoglobin 8.46f Mural Rexine wallcovering
Nitrocellulose-coated fabric, stencil-printed
Crystallographer: John Kendrew
Designer: Charles C. Garnier
Pattern no: CR9; 1 colourway
Collections: V&A Circ.52-1968
Applications: Limited production in 1951. Used on walls in Telephone Recess at the Exhibition of Science.

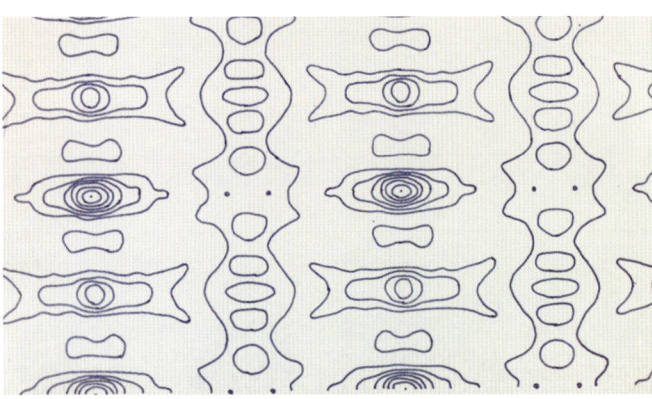

Myoglobin 8.46c diagram

FPG 30
Myoglobin 8.46c Vynide upholstery fabric
Nitrocellulose-coated fabric, stencil-printed
Crystallographer: John Kendrew
Designer: Charles C. Garnier
Pattern no: CR6; *repeat*: 3.2 x 9.5cm; 1 colourway
Collections: V&A Circ.52-1968
Souvenir Book, p.7
Limited production in 1951. Shown in Land Travelling Exhibition.

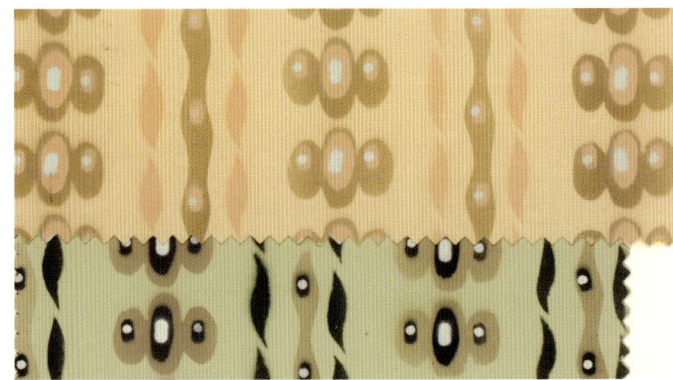

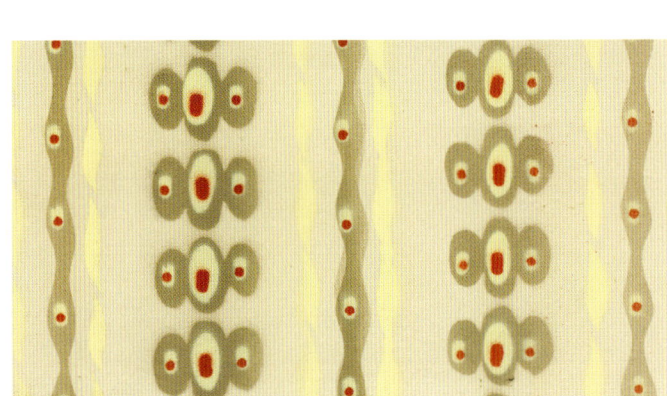

FPG 32
Myoglobin 8.46g Vynide upholstery fabric
Nitrocellulose-coated fabric, stencil-printed
Crystallographer: John Kendrew
Designer: Charles C. Garnier
Pattern no: CR1; *repeat*: 10.7 x 7cm; 4 colourways
Collections: V&A Circ.52-1968; Circ.56-1968+A
Souvenir Book, p.7 and p.16
Limited production in 1951. Used on benches in cinema foyer at Exhibition of Science. Shown in Land Travelling Exhibition.

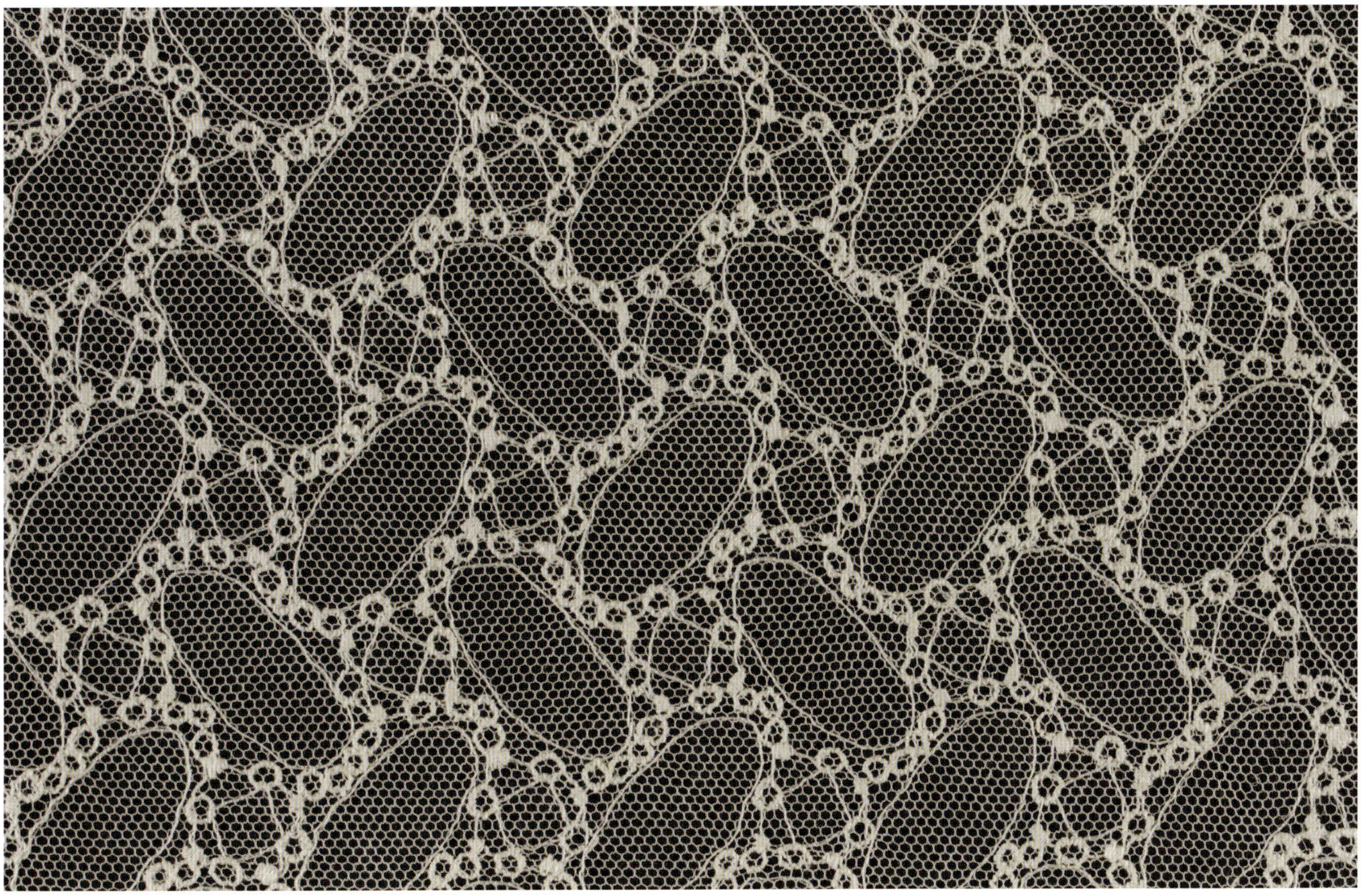

FPG 18 - Apophyllite 8.30 lace, designed by H. Webster for A.C. Gill

FPG 22 - Polythene 8.59 lace, designed by H. Webster for A.C. Gill

ARNOLD LEVER
FPG code: A515

Textile designer and manufacturer based in London. Arnold Lever made his name during the 1940s designing dress fabrics and scarves for Jacqmar, an innovative company founded in 1932 by Jack and Mary Lyons. Lever established his own company in 1947, designing and producing printed dress fabrics for the couture market and leading department stores.

Lever agreed to participate in the FPG on 18/8/1949 and proposed a range of crepe de chine suitable for Liberty's or John Lewis. At the first meeting of the Textiles Subcommittee, 31/1/1950, he was described as a producer of 'high fashion materials in all fibres (rayon, silk, cotton etc.).' Two of the four designs he presented at the FPG meeting on 13/7/1950 were considered too free, so at Lever's initiative they were withdrawn.

Archives: AAD 1977/3/217-22; DCA 5384-1, 5384-3, 5396

FPG 33
Afwillite dress fabric ('Fibre')
Screen-printed silk or fine filament acetate rayon crepe
Crystallographer: Helen Megaw
Designer: Arnold Lever
Pattern nos: 472, 473; 2 colourways
Collections: SM 1976-644/42
Skinner's Record, p.475
In production in 1951 under the name 'Fibre'. Pattern derived from a Fourier contour map of Afwillite but does not match any recorded diagrams; possibly a free interpretation of 8.44. Lever sent HM a length of silk in beige, 22/5/1951, which she had made into blouse to wear at the International Congress of Crystallography, Stockholm, June 1951. 'I believe the design you wanted for yourself was one of the smaller cuttings enclosed of "Fibre". Will you please examine them under your microscope and let me know which one you like best and I shall be delighted to send you whatever length you need.' (Letter from Arnold Lever to HM, 17/5/1951).

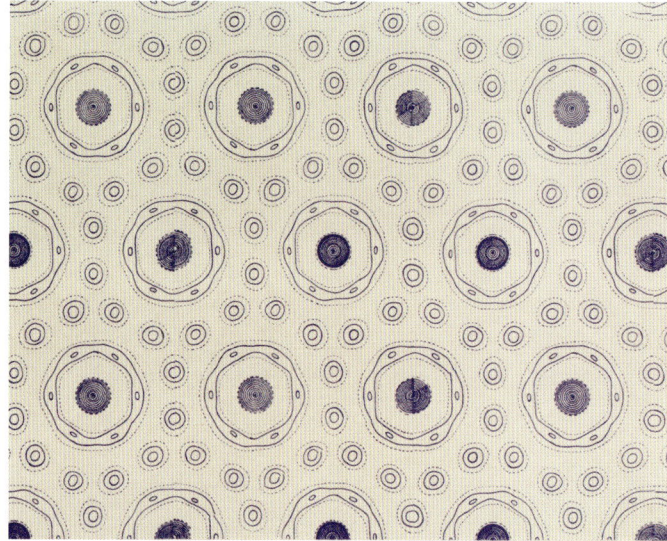

Haemoglobin 8.26 diagram

FPG 34
Haemoglobin 8.26 dress fabric ('Flying Saucers')
Screen-printed silk or fine filament acetate rayon crepe
Crystallographer: Max Perutz
Designer: Arnold Lever
Pattern nos: 1310, 1311, 1312, 1313; *repeat*: 38 x 38cm (approx.); 4 colourways
Collections: V&A Circ.76-1968 [missing]; SM 1976-644/41
Souvenir Book, p.2; *Architectural Review*, p.238; *British Textiles*, p.54
In production in 1951 under the name 'Flying Saucers'. Shown in FPG displays in Regatta Restaurant and Land Travelling Exhibition.

JOHN LINE & SONS
FPG code: A514

Wallpaper manufacturers and merchants based in London. Grew out of a furniture company in Reading established by cabinetmaker John Line and his three sons. By the 1880s they had branched out into wallpaper, initially as wholesalers, subsequently as producers as well, relocating to London in 1892. John Line's collections were diverse, ranging from luxurious block-printed and stencilled patterns, to inexpensive machine prints. At the Festival of Britain they used a new technique, hand screen printing. In 1958 John Line merged with Shand Kydd. Both companies were absorbed into the Wall Paper Manufacturers Ltd. in 1961. The John Line brand disappeared later that decade after the Monopolies Commission decreed that the WPM should be broken up.

John Line were approached about the FPG on 18/8/1949 and became enthusiastic participants. Henry G. Dowling, Chief Decorative Adviser, attended the first FPG meeting, 16/12/1949. 'You have unique opportunities to influence manufacturers and designers in the right direction, and I would like again to congratulate you on all the good work you are doing, and to assure you that you can certainly count on our wholehearted support.' (Letter from Henry Dowling to MHT, 29/1/1951).

John Line's staff designers, led by William Odell, developed numerous crystal structure designs. Only four were approved, but at least two others were developed. These wallpapers were used extensively in the Regatta Restaurant and Exhibition of Science and were also shown in the Dome of Discovery. 'We made some 35 designs here and actually produced four in several colourings. These colourings were made to definite requirements and were for exhibition purposes only, the designs themselves being in scale to suit the spaces on which they were eventually to be used; but we have other designs which, from our point of view, might be even better than those selected, and the colourings of those which were used could be made much more saleable; and it is our intention to re-colour some of the Exhibition designs, as well as to produce them to a much smaller and more useable scale. / From this you will gather that we have found the experiment most interesting, and the samples we have shown have created great interest.' (Letter from Henry G. Dowling to MHT, 12/10/1951). HM was also very positive: 'I should very much like to have some of them to show off to people who are interested in the idea, and might put up some bits on the wall of my room in the laboratory.' (Letter from HM to Henry Dowling, 29/4/1951).

Archives: AAD 1977/3/580; DCA 5384-2, 5384-3, 5384-4, 5396

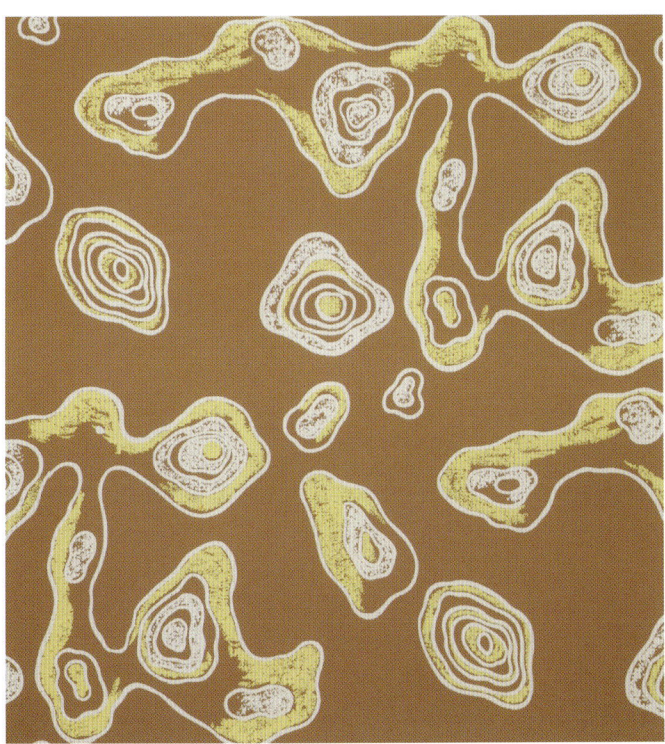

FPG 35
Afwillite 8.45 wallpaper
Hand screen-printed wallpaper
Crystallographer: Helen Megaw
Designer: William J. Odell
Repeat: 53.3 x 53.3cm; 2 colourways
Collections: V&A E.885-1978; E.2345-1980; SM 1976-644/48
Souvenir Book, p.6 and p.16; *Architectural Review*, p.239
In production in 1951. Used in Regatta Restaurant in dark green (5 pieces); also shown in FPG display. Used extensively on display stands in Exhibition of Science.

FPG 36
Boric Acid 8.34 wallpaper
Hand screen-printed wallpaper
Crystallographer: Helen Megaw
Designer: William J. Odell
Repeat: 53.3 x 46cm; 3 colourways
Collections: V&A E.886-1978; E.2346-1980; MAG 1986.161; SM 1976-644/46 (artwork)
Souvenir Book, p.9
In production in 1951.

FPG 37
Insulin 8.25 wallpaper
Hand screen-printed wallpaper
Crystallographer: Dorothy Hodgkin
Designer: Robert Sevant
Repeat: 26.7 x 26.7cm; 2 colourways
Collections: V&A E.888-1978; E.2342-1980; MCAG 1986.151; SM 1976-644/51
Souvenir Book, p.5: 'Sevant converts the hexagonal repeat into a square one and drops half the original in doing so.' *Architectural Review,* p.238
In production in 1951. Used in Cinema Foyer at the Exhibition of Science. Shown in Land Travelling Exhibition.

FPG 38

FPG 38
Insulin 8.27 wallpaper
Hand screen-printed wallpaper
Crystallographer: Dorothy Hodgkin
Designer: William J. Odell
Repeat: 53.3. x 53.3cm; 1 colourway
Collections: V&A E.884-1978; E.2344-1980; SM 1976-644/50
Souvenir Book, p.4: 'The change from the hexagonal grid of the diagram to a square repeat was radical, but despite the bold treatment, character is not lost.' *Architectural Review,* p.238
In production in 1951. Used on two walls in Regatta Restaurant (11 pieces).

Insulin 8.25 diagram   *Insulin 8.27 diagram*

FPG 39
Mica 8.35 [2] wallpaper
Hand screen-printed wallpaper
Crystallographers: W.W. Jackson / J. West
Banham and Hillier (1976), pl. II
Probably sample production only.

FPG 40
Nylon 8.54c wallpaper ('Cherwell')
Hand screen-printed wallpaper
Crystallographer: Charles William Bunn
Designer: William J. Odell
Collections: V&A E.2324-1980
Sample production only under the name 'Cherwell'. Not shown at the Festival, but created as part of the FPG scheme.

LINOLEUM MANUFACTURING COMPANY
Tradename: Staines Linoleum
FPG code: A530

Linoleum manufacturers based in Staines, Surrey. The Linoleum Manufacturing Company was established by Frederick Walton, the inventor of linoleum, in 1864. In 1929 it merged with Barry, Ostlere & Shepherd Ltd. of Kirkcaldy to become part of Barry & Staines Linoleum Ltd., but continued to operate autonomously, specialising in inlaid linoleum. Manufacturing continued at Staines until 1969.

E.N. Barran, Director, was approached about the FPG in January 1950 at the recommendation of Jas. Williamson & Son, who specialised in printed linoleum. Originally it was thought that these two firms might supply flooring for the Regatta Restaurant. Design drawings were sent by D.C. Beese, Works Manager, on 29/6/1950, but Misha Black decided to use carpet instead. Later it was proposed that these designs might be adapted for the Exhibition of Science, but this offer was subsequently withdrawn. 'I share your disappointment and agree with you heartily that the architect should have decided much earlier that he could not use linoleum. / I do not however quite see why you necessarily withdraw your offer to Mr Brian Peake for the Science Museum. I seem to remember that it was a question of machine-made in a larger quantity or handmade if only small. Cannot you revert to the handmade? / In any case I take it that you will prepare handmade samples to occupy your place in the Group's joint display in the foyer of the restaurant... / I should like, if you will let me, to assist you in the further development of your designs so that you will have something really commanding in the display.' (Letter from MHT to E.N. Barran, 15 September 1950).

The company announced their resignation from the Group on 6/11/1950, but expressed a desire for their designer, E.H. Tee, to continue to receive advice from MHT. The resulting design, Insulin 8.27, was reproduced anonymously in FPG publications at their insistence, credited to 'a linoleum manufacturing company'. 'Their contribution in the end amounted to a sketch for a design in stencil inlaid linoleum, which they did in duplicate, one in the Group display in the Regatta Restaurant, South Bank Exhibition, and the other in the Group display in the Land Traveller. The sketch was polished up to look like a piece of linoleum superficially. I must accept some responsibility for the design, because I suggested that they took this rather florid design and coloured it in the style of their imitation turkey carpet linoleum.' (Letter from MHT to J.P. McCrum, COID, Glasgow, 8/10/1951).
Archives: DCA 5384-1, 5384-2, 5384-3, 5384-4, 5396

FPG 41
Insulin 8.27 linoleum
Design for stencil-inlay linoleum
Crystallographer: Dorothy Hodgkin
Designer: E.H. Tee
Repeat: 46 x 91.5cm; 2 colourways
Collections: V&A Circ.43-1968 [missing]
Souvenir Book, Cover and p.4: 'The bright colours are for those who like Turkey carpet – from an industry that has produced imitations of every other conceivable floor-covering. The grain here is not imitation carpet-tufting, but the proper result from an ingenious mechanical process.'
Prototype only, not developed beyond artwork.

FPG 62 - Insulin 8.25 tie fabric, designed by George Reynolds for Vanners & Fennell

FPG61 - Haemoglobin 8.26 tie fabric, designed by George Reynolds for Vanners & Fennell

LONDON TYPOGRAPHICAL DESIGNERS

FPG code: A526

Graphic designers based in London. London Typographical Designers was formed in 1945 by typographers Leon French and William Morgan, artist Beric Young and master printer Oliver Burridge, who had all worked together at the Ministry of Information during the war. George A. Thompson joined in 1948 and Ken Sewell in 1953. The company still operates today as LTD Design Consultants.

London Typographical Designers accepted a late invitation to join the FPG on 12/1/1950. They designed graphics for the Regatta Restaurant and the FPG display. They were also responsible for publishing *The Souvenir Book of Crystal Design*s. On 10/7/1952 they asked HM to supply crystal structure diagrams of china clay, ball clay, flint and marble for a commission from a tile-making company.

Archives: AAD 1977/3/278, 280-285; DCA 5384-3, 5384-4, 5396; University of Reading Archives

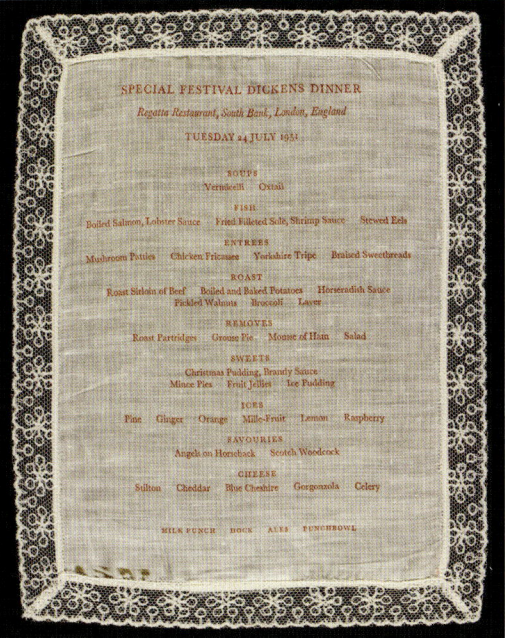

FPG 43

Hydrargillite 8.33 menu for Festival Dickens Dinner

Printed linen trimmed with machine-embroidered cotton lace

Crystallographer: Helen Megaw

Designers: London Typographical Designers

Collections: V&A Circ.65-1968

Special menu for event at Regatta Restaurant, 24/7/1951, trimmed with lace by A.C. Gill: 'the lace menu for the Dickens Dinner... obtained world-wide press mention with several photographic reproductions appearing in several newspapers and magazines at home and abroad.' (Letter from Leon French to MHT, 12/10/1951).

FPG 42

Hydrargillite 8.33 menu and wine list covers

Embossed and printed nitrocellulose-coated Rexine (menus) and leather (wine lists)

Crystallographer: Helen Megaw

Designers: London Typographical Designers

Souvenir Book, p.15

Used in the Regatta Restaurant by the caterers, Hagenbachs. Same motif used on signage in restaurant.

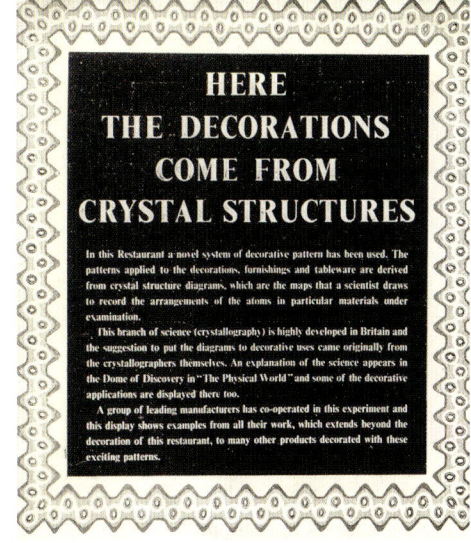

FPG 44

Polythene 8.59c graphics panel

Printed paper with decorative border

Crystallographer: Charles William Bunn

Designers: London Typographical Designers

Souvenir Book, p.14

Featured in FPG display in Regatta Restaurant.

OLD BLEACH LINEN COMPANY
FPG code: A513

Textile manufacturers based in Randalstown, Northern Ireland. Established in 1864, Old Bleach Linen Company specialised in jacquard-woven table linen and furnishing fabrics. During the 1930s the mill employed over 1000 people and undertook major contracts for hotels and shipping companies. Screen-printed textiles were produced from 1930s; tea towels became increasingly important after World War II. Some cotton fabrics were produced in addition to linen. The factory closed in 1980.

Old Bleach were invited to join the FPG on 17/11/1949. Although initially sceptical because of the amount of time required to develop jacquard-woven designs, Norman Webb, Director, agreed to participate and attended the first Textile Subcommittee meeting, 31/1/1950. Six samples and three sketches were presented, 11/7/1950. The original aim was to produce table linen and curtains for the Regatta Restaurant, but this proved impossible due to decision-making delays. Some furnishing fabrics were produced, but they were only ever woven in fairly small quantities. Samples were shown in the Dome of Discovery.

Archives: AAD 1977/3/192, 195, 196, 351-369; DCA 5384-1, 5384-3, 5384-4

FPG 46
China Clay 8.6 furnishing fabric
Jacquard-woven linen or cotton
Crystallographer: G.W. Brindley
Pattern no: Z1404, CW282/128; w.127cm
Collections: WAG T.1992.35.1-2; SM 1976-644/28; V&A T.117, 182-1992
Limited production in 1951; shown at British Industries Fair. Originally produced in linen, but woven in cotton for curtains in Lawrence Room at Girton College, Cambridge at HM's instigation. Wrong material sent initially, but order rectified, 17/10/1951. (Top image)

FPG 45
China Clay 8.6 table linen
Jacquard-woven unbleached linen damask with white rayon weft
Crystallographer: G.W. Brindley
Pattern no: Z1394; *repeat*: 3.2 x 6.4cm
Collections: V&A Circ.71-1968; SM 1976-644/33
Souvenir Book, p.13; *British Textiles*, p.50; *Queen*
Prototype table mats and napkins were photographed but never manufactured. Samples shown in FPG display in Regatta Restaurant.

FPG 47
Orthoclase 8.29 furnishing fabric
Jacquard-woven linen
Crystallographer: W.H. Taylor
w.127cm; *repeat*: 5.8 x 9.5cm; 1 colourway
Collections: V&A Circ.65-1952+A; SM 1976-644/32 + 52
Souvenir Book, p.9; *British Textiles*, p.50; *Homes and Gardens*, p.34; *Architectural Review* (December 1951), p.358
Limited production in 1951; shown at British Industries Fair.

FPG 48
Hydrargillite 8.33 furnishing fabric
Jacquard-woven linen or cotton
Crystallographer: Helen Megaw
w.127cm; *repeat*: 5.8 x 9.6cm; 4 colourways
Collections: V&A Circ.64-1952+A, T.367-1977; SM 1976-644/29, 40 + 52
Souvenir Book, p.9: 'Old Bleach and Warerite have both copied the hydrargillite diagram faithfully, but have altered the spacing to suite the two different applications.' *Homes and Gardens*, p.34; *Design*, p.38 (advertisement): 'For modern homes, a design as modern as to-day! Inspired by the atomic structure of aluminium hydroxide. Woven in durable, colour-fast Irish linen.' *Architectural Review*, December 1951, p.358
Limited production in 1951; shown at British Industries Fair.

R.H. & S.L. PLANT

Tradename: Tuscan China
FPG code: A510

Ceramics manufacturers specialising in bone china, based in Longton, Stoke-on-Trent. Established as R.H. Plant & Co. in 1881, but renamed R.H. & S.L. Plant in 1898 after the founder's brother joined the firm. Formed partnership with Susie Cooper in 1958; both firms taken over by Wedgwood Group in 1966. Brand renamed Royal Tuscan in 1971.

Initially approached about the FPG by Professor Baker; invited to participate in October 1950. Hagenbachs asked if they could produce cake stands or fruit dishes for the restaurant, but Plant's said they could only make a token contribution due to shortage of time. The resulting tea service, 'Festival', was designed by one of Baker's former students, Hazel Thumpston, who had participated in a project to create crystal structure designs at the RCA.
Archives: AAD 1977/3/225, 332-348, 585; DCA 5384-2, 5384-3, 5384-4, 5396

FPG 49
'Festival' tea service
Bone china, printed in overglaze enamels, hand painted enamel dots, gilded rim and banding
Designer: Hazel Thumpston
Side plate: d.16.5cm; *dessert plate*: d.22.8cm; *dinner plate*: d.27.3cm
Collections: V&A Circ.39-1968 (dinner plate)
Souvenir Book, p.14: 'A general essay on the Crystal Patterns theme.' Prototypes only, although the manufacturer expressed a desire to produce the design at a later stage. 'I...am sending you a sample of the pattern which we regard as being very good. It has been admired generally, and although very difficult to produce commercially, it will be possible, we hope, ultimately to get it on the market.' (Letter from

H.J. Plant to HM, 7/5/1951). Samples were shown in FPG displays in Regatta Restaurant and Land Travelling Exhibition.

This is the only pattern not based on an identifiable crystal structure diagram. According to Professor Baker, the pattern was derived from a lightgraph of Bismuth Oxychloride. H.J. Plant, Managing Director, claimed that: 'The chosen pattern is a combination of the whorl motif from the Adenine Hydrochloride (diagram 8.48) with a section of the Hydrargillite crystal (diagram 8.33)'; but MHT noted that: 'conversations with Miss Thumpston made it clear that these ascriptions are ex post facto. What she really did was a free essay on the FPG theme.' HM noted on a photograph: 'Based on Bernal chart used for interpreting oscillation photographs (X-ray) of crystals, and not on any structure.'

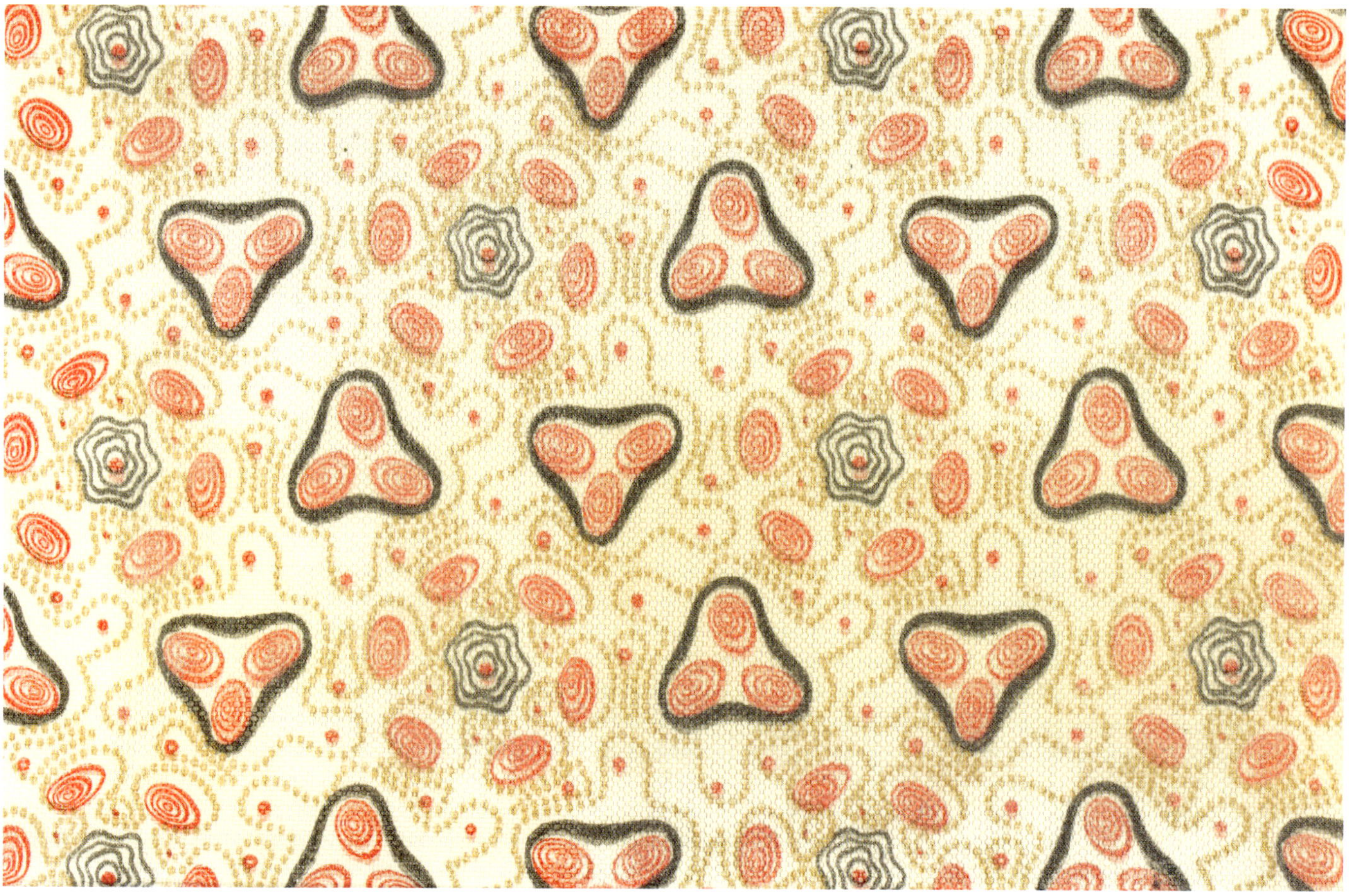

FPG 29 - Insulin 8.24 Mural Rexine, designed by Charles C. Garnier for ICI Leathercloth

FPG 31 - Myoglobin 8.46f Mural Rexine, designed by Charles C. Garnier for ICI Leathercloth

SPICERS

FPG code: A527

Paper makers and stationery manufacturers based at Sawston, Cambridgeshire. Founded in 1796 by John Edward Spicer; run by his four sons during the 19th century when it became known as Spicer Brothers. Merged with James Spicer & Sons to become Spicers Ltd. in 1922; remained a family business until after World War II; public company from 1951. Taken over by Reed in 1963, but became independent again in 1988 after a management buyout. In 1993 Spicers was sold to a paper and packaging company called David S. Smith Holdings plc. Still operational today.

Spicers did not become involved in the FPG until late 1950, but were enthusiastic participants, producing 'fancy' wrapping paper and gift boxes. 'We hope to have the rollers available in some 14 days and the first printing pulls inside three weeks". (Letter from Spicers to COID, 19/2/1951). D.C. Spicer, Director, was keen to see the initiative continued after the Festivalalthough he queried whether the creative momentum could be sustained if the COID were not directly involved.

In 1959 HM wrote to Spicers asking if they would produce some new crystal structure designs for the forthcoming International Union of Crystallography conference in Cambridge in 1960: 'I have a lot of designs in my files which have not been used and which would be very attractive.' (Letter from HM to C. Goody, 26/8/1959). She visited their factory, but in the end Spicers decided not to proceed.

Archives: AAD 1977/3/189, 351-369; DCA 5384-3, 5384-4, 5396

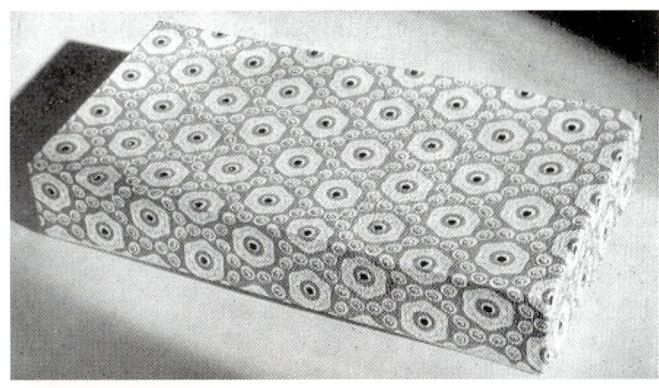

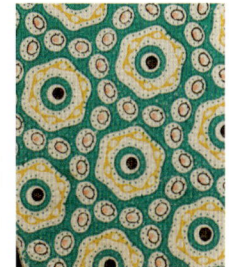

FPG 50
Haemoglobin 8.26 decorative paper
Roller-printed fancy wrapping paper and boxes
Crystallographer: Max Perutz
Designer: W. Farquhar
w.51 x l.76cm; 2 colourways
Collections: SM 1993-421; Girton College, Cambridge
Souvenir Book, p.15
'Fancy wrapping paper' and 'fancy boxes' were in production in 1951, obtainable through stationers. Box shown in FPG display in Regatta Restaurant; paper shown in Land Travelling Exhibition.

STEVENS & WILLIAMS

Tradename: Royal Brierley Crystal
FPG code: A507

Glass manufacturers specialising in lead crystal, based in Stourbridge, West Midlands. Stevens & Williams was founded in 1847 when William Stevens and Samuel Cox Williams took over their father-in-law's glassworks at Briar Lea Hill. During the late 19th century they were one of the most innovative ornamental glassmakers in the country. Later they focused increasingly on cut lead crystal; the name Royal Brierley Crystal was adopted in 1931. The company changed hands several times from 1998 onwards. Royal Brierley Crystal brand is now allied with Dartington Crystal, but production ceased at Brierley Hill in 2002.

H.S. Williams-Thomas, Chairman, was asked to supply glassware for the Regatta Restaurant in October 1949. Major H. McEwan attended the first FPG meeting, 16/12/1949. Their original design was rejected by the caterers on 28/9/1950; Hagenbachs thought it would get stolen as it looked too much like a souvenir. An all-over pattern was requested instead. A mould-blown water set was also discussed, but later abandoned. In the end their contribution was limited to a prototype wineglass by their in-house designer, Sam Thompson. 'Unfortunately, we are so much handicapped in the home market by Board of Trade limitations that we could not co-operate to the extent we may have done under normal conditions. I feel also that glass as a medium has not the same opportunity of making use of the crystalline theme which could be used to such advantage in other materials.' (Letter from H.S. Williams-Thomas to MHT, 1/10/1951).

Archives: DCA 5384-1, 5384-2, 5384-3, 5384-4, 5396

FPG 51
China Clay 8.6 wineglass
Lead crystal, printed in white enamel
Crystallographer: G.W. Brindley
Designer: Sam W. Thompson
Souvenir Book, p.13; *Homes and Gardens*, p.35
Prototype only. Shown in FPG display in Regatta Restaurant.

James Templeton & Co.

FPG code: A508

Carpet manufacturers based in Glasgow. In 1839 shawl manufacturer James Templeton patented a new chenille-making process which was successfully applied to carpets, leading to the formation of James Templeton & Co. in 1843. The company grew rapidly, also producing Brussels and Wilton carpets. By 1921 Templeton's had six factories in Glasgow and one in Stirling. In 1983 the company merged with two other firms to form Stoddard Carpets Ltd. Stoddards went into receivership in 2005.

Templeton's were approached about the FPG on 17/8/1949 with a view to producing carpet for the Regatta Restaurant. Managing director John Anderson and chief designer Hugh McKenna both responded positively: 'They [ie. crystal structure diagrams] are most interesting indeed when examined at closer hand and offer almost limitless possibilities in the use of colour. We shall certainly carry out some experimental work soon and I feel sure that with all round co-operation we could obtain very satisfactory results from this project.' (Letter from Hugh McKenna to MHT, 12/10/1949). At the first FPG meeting McKenna showed designs based on copper, quartz, durangite and mica. 'These were most successful, and Dr. Megaw said that with one exception, where some atoms had been marked and not others, she considered the patterns sufficiently faithful representations of the crystal diagrams. The use of colour was particularly effective, and showed how the choice could be entirely free, provided it was followed consistently. It was thought that the slightly irregular forms, such as those provided by insulin, were more attractive than the strictly geometrical patterns.' (Minutes of first FPG meeting, 16/12/1949). Artwork for these designs survives in the Stoddard Carpets Archive.

Templeton's designers create a large number of designs, five of which were developed into carpets. Artwork for two other patterns based on Insulin 8.27 is recorded in an archive photograph (AAD 1977/3/581). The carpets were used extensively in the Regatta Restaurant and at the Exhibition of Science, including in the reception suite. However, in spite of a marketing campaign, orders were not forthcoming. 'We illustrated them in colour in our mid-summer House Magazine; we had carpets on display in our showroom here and in London. Samples were sent to all our overseas warehouses and you may be aware Liberty's have a display in two of their Regent Street windows which give prominence to two of our carpet designs. I regret to say that we have not received a single trade order for any one of these designs, and meanwhile it would appear that such extra stock as we have provided, may have to be offered at a substantial discount before they are ultimately disposed of. In the circumstances, I think you will agree that there would not appear to be any purpose in our continuing to pursue this interesting experiment.' (Letter from John Anderson to MHT, 3/10/1951).

Archives: AAD 1977/3/581-582; DCA 5384-2, 5384-3, 5384-4, 5396. Stoddard Carpets Archive (in the process of being acquired by University of Glasgow)

FPG 52
Insulin 8.25 carpet
Machine-woven carpet
Crystallographer: Dorothy Hodgkin
Designer: G. Brown
Pattern no: 0837; *repeat*: 37 x 68.5cm; 1 colourway
Collections: V&A Circ.50-1968
Souvenir Book, p.5
Limited production in 1951.

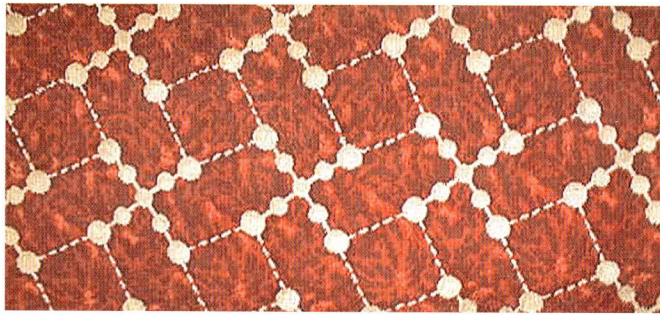

FPG 53
Pentaerythritol 8.18 carpet
Machine-woven carpet
Crystallographer: Gordon Cox
Designer: J.M. McCreery
Pattern no: 0840
Collections: Chemistry Department, University of Leeds
British Textiles, p.55; *Architectural Review* (December 1951), p.358
Limited production in 1951. Used in Exhibition of Science on computer machine platform in Stop Press section; this sample subsequently re-used in the office of Professor Gordon Cox, Chemistry Department, University of Leeds. Sample shown at Liberty's, August 1951.

FPG 54
Perovskite 8.1 carpet
Machine-woven carpet
Crystallographer: Helen Megaw
Designer: L. Halliday
Collections: V&A Circ.51-1968
British Textiles, p.55; *Architectural Review* (December 1951), p.358
Limited production in 1951. Used in Exhibition of Science (section designed by Eric Mansfield).

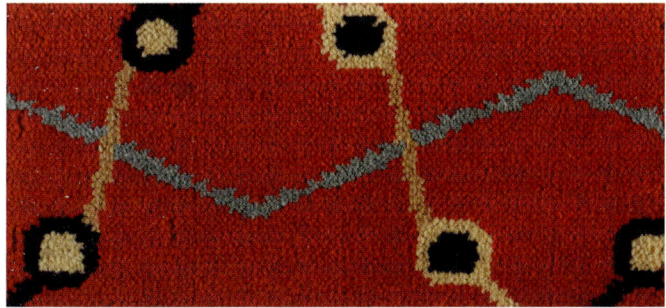

FPG 55
Quartz 8.10 carpet
Machine-woven carpet
Crystallographers: W.H. Bragg / Lawrence Bragg
Designer: Robert Anderson
Pattern no: 0838; *repeat*: 21.5 x 26.7cm
Collections: V&A Circ.49-1968
Souvenir Book, p.11: 'One set of dots has been omitted in the adaptation. The diagrams are specially suited to making patterns for carpets, because the strongly marked lattice patterns measure out the space in a room and give it scale.'
Limited production in 1951. Used in the Exhibition of Science. Shown in Land Travelling Exhibition.

FPG 56
Resorcinol 8.17 carpet
Machine-woven carpet
Crystallographer: John Monteath Robertson
Designer: Robert Anderson
Pattern no: 0846; *repeat*: 17.8 x 11.5cm; 4 colourways
Collections: V&A Circ.48-1968
Souvenir Book, p.10: 'The small dots, representing atoms, have been omitted, and the lozenges have been straightened to suit the weaving.' Limited production in 1951. Used in Regatta Restaurant in two colourways: purple madder and attic rose in main restaurant (925yds); dark green and dark purple in balcony restaurant (175yds). Shown in FPG display in Regatta Restaurant.

VANNERS & FENNELL
FPG code: A522

Textile manufacturers specialising in jacquard-woven silks for ties, cravats, umbrellas, banners, facings and clerical garments, based in Sudbury, Suffolk. Vanners & Fennell Bros. Ltd. were formed in 1886 from the merger of two silk manufacturers: Vanners of Sudbury and Fennell Brothers of Haverhill. The Vanners were French Huguenot silk weavers active in Spitalfields during the 18th century, who later moved to East Anglia. In 1924, after acquiring the Sudbury Silk Weaving Company, E.W.G. Kipling and Anthony Rowland became directors. After World War II tie fabrics became the main focus, mainly sold to major tie manufacturers such as Welch Margetson & Co. Vanners & Fennell acquired the rights to the Welch Margetson name in the 1970s after the firm went into liquidation. Still operational today, specialising in luxury silk tie fabrics woven at their Gregory Street mill in Sudbury, Vanners (as they are now known) are part of a larger company called Silk Industries Ltd.

Vanners & Fennell were approached by MHT on 7/6/1950. Gordon Kipling, Technical Director, attended an FPG meeting on 13/7/1950, but Bernard Rowland, Sales Director, led the project after that. The company entered enthusiastically into the project, producing seven tie fabrics, all recorded in the company's pattern book on 21/7/1950. Rowland, who was responsible for the American market, was credited as designer in contemporary publicity, but the range was actually designed by George Reynolds, the firm's chief designer.

FPG ties manufactured by Welch Margetson (retail price 21s) were popular with cyrstallographers, much to HM's satisfaction. 'I am writing to tell you of the widespread interest I have met with among crystallographers in your ties. I was at an international conference at Stockholm last month, and many of the Cambridge people were wearing them. I also noticed Dr Patterson, the American author of the Patterson diagram, wearing the Patterson diagram of insulin, to his own very obvious pleasure and everybody else's. A lot of people were making inquiries about

the ties, and a stream of American visitors coming back though this country eventually cleared out the whole stock of Shepherd's in Trinity St. [Cambridge], so that late comers had to go away disappointed... / A good many ties of very varied patterns and colouring have been bought by my colleagues in the Cavendish Laboratory, and give pleasure every time I see them..' (Letter from HM to Bernard Rowland, 6/8/1951).

Fabrics were shown in FPG displays in Dome of Discovery and Regatta Restaurant. Haemoglobin ties were produced for FPG members; the manufacturers sent HM a Perovskite scarf. She wanted to buy a length of Aluminium Hydroxide, but was told that all the fabric had been cut up by a wholesale house. 'I am disappointed that there is not going to be any chance to buy these tie-silks for dress materials, with tie-silks so much in fashion for the purpose; I think they might have been extremely smart. However, they would probably also have been extremely expensive.' (Letter from HM to Bernard Rowland, 3/6/1951). In May 1951 she sent the company some new crystal structure diagrams in the hope that they might produce some more fabrics. 'I shall be very glad to have cuttings of any other patterns you produce. Curiously enough, just after I had posted my last letter to you including a diagram of the spinel structure, I was asked by Professor Jeffreys whether there were any spinel ties; he wants one because it is his theory that the inside of the earth consists of spinel!' (op. cit.)

Bernard Rowland downplayed the commercial success of the ties, whilst acknowledging the value of the project: 'The volume of business was comparatively small, due to the fact that in order to produce rich designs of this nature the market was very limited owing to the high price of the resultant article.... / We certainly gained some advertisement from the venture... and it was a great pleasure to work with all the other members of the Group.' (Letter from Bernard Rowland to MHT, 12/10/1951). However, according to Andrew Henry, current Managing Director of Vanners, all seven designs were very successful; over 600 ties were sold in the US market alone. On 11/1/1954 Vanners

& Fennell told HM they wanted to produce some more fabrics. She sent them some patterns and offered to waive fees in return for samples. In the end they decided not to go ahead, mainly because of concerns about copyright, but also because they were told that 'all over' patterns were no longer in fashion. However, they reissued several existing designs and offered HM a length of China Clay fabric by way of recompense which she used for a jacket. On 26/8/1959 HM asked if they would put some fabrics back into production to make ties for the International Union of Crystallography conference in Cambridge the following year (15-24 August 1960). Welch Margetson agreed to produce China Clay ties on the basis of a substantial order from the Cambridge shop Shepherd's.

Archives: AAD 1977/3/182, 185-187, 241, 309-310, 313-324, 367; DCA 5384-2, 5384-3, 5384-4, 5396; VAN

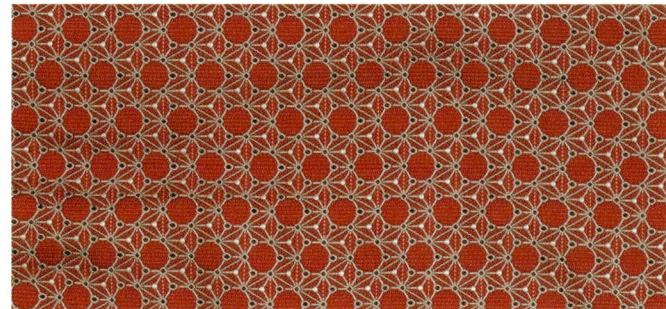

FPG 59
China Clay 8.6 tie fabric
Jacquard-woven silk
Crystallographer: G.W. Brindley
Designer: George Reynolds
Pattern nos: F8; 74495; 62110; w.76cm; *repeat:* 1.2 x 2.2cm; 2 colourways
Collections: V&A T.368-1977 (jacket); T.446F-1977; VAN
Souvenir Book, p.3 and p.13
In production in 1951; reintroduced 1954 and 1960. Ties manufactured by Welch Margetson.

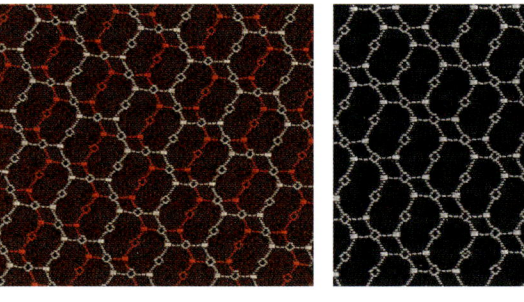

FPG 60
Clay Minerals 8.55 tie fabric
Jacquard-woven silk
Crystallographer: G.W. Brindley
Designer: George Reynolds
Pattern nos: F5 and F6; w.76cm; 3 colourways
Collections: V&A T.446F-1977; VAN
In production in 1951. Ties manufactured by Welch Margetson.

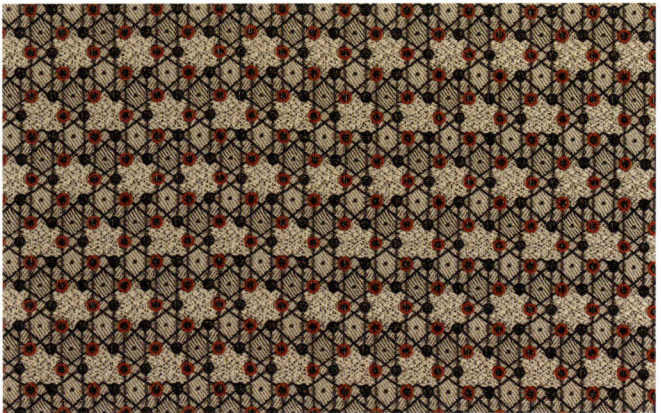

FPG 57
Aluminium Hydroxide 8.4 tie fabric
Jacquard-woven silk
Crystallographer: Helen Megaw
Designer: George Reynolds
Pattern no: F4; w.76cm; 3 colourways
Collections: V&A T.446F-1977; VAN
In production in 1951. Ties manufactured by Welch Margetson

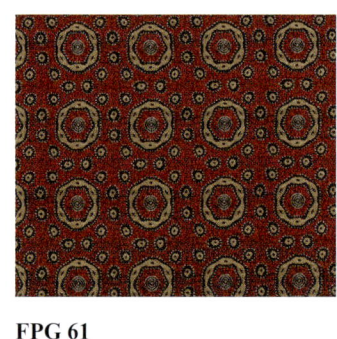

FPG 61
Haemoglobin 8.26 tie fabric
Jacquard-woven silk
Crystallographer: Max Perutz
Designer: George Reynolds
Pattern nos: F1; 74666; w.76cm; *repeat:* 4 x 4cm; 3 colourways
Collections: V&A Circ.72-1968; T.446F-1977; VAN
Souvenir Book, p.3: 'the most successful adaptation of the haemoglobin diagram, though the triangular framework has been altered to a square.'
Architectural Review, p.238; *British Textiles,* p.54
In production in 1951; reintroduced 1954. Ties manufactured by Welch Margetson; distributed to FPG members.

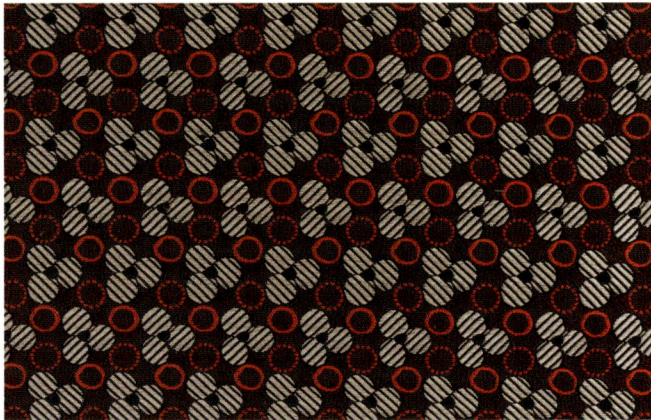

FPG 58
Chalk 8.28 tie fabric
Jacquard-woven silk
Crystal structure: 'Common knowledge'
Designer: George Reynolds
Pattern no: F3; w.76cm; *repeat:* 5 x 4cm; 1 colourway
Collections: V&A T.446F-1977; VAN
In production in 1951. Ties manufactured by Welch Margetson.

FPG 62
Insulin 8.25 tie fabric
Jacquard-woven silk
Crystallographer: Dorothy Hodgkin
Designer: George Reynolds
Pattern nos: F2; 74969, 74970; w.76cm; *repeat*: 7.5 x 4.5cm; 2 colourways
Collections: V&A T446F-1977; VAN
In production in 1951; reintroduced 1954. Ties manufactured by Welch Margetson.

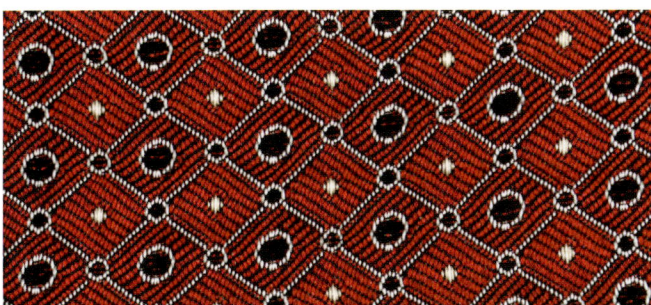

FPG 63
Perovskite 8.1 tie fabric
Jacquard-woven silk
Crystallographer: Helen Megaw
Designer: George Reynolds
Pattern no: F7; w.76cm; 2 colourways
Collections: V&A T.446F-1977; VAN
Applications: In production in 1951. Ties manufactured by Welch Margetson. HM was given a scarf in April 1951.

VERNON INDUSTRIES
Tradename: Oxvar
FPG code: A518

Furniture manufacturers based in Liverpool. Vernon Industries were owned by Vernon Sangster, co-founder of Vernon Pools. Specialising in industrially-manufactured furniture made of metal and wood, the company produced the 'House Proud' range of planned kitchen equipment, as well as office furniture and storage units for large-scale contract use.

The company was recommended for inclusion in the FPG by MHT's colleague, S.D. Cooke: 'I had a talk with Mr. Skelton of Vernon Industries some days ago on the subject of their new method of applying wood grain finish to metal sheets, synthetic boards, etc., and I think they would be very interested in your crystallography project. / In view of the fact that they have a large contract for steel lockers for the National Coal Board, and also they are carrying out some important contracts for office furniture for His Majesty's Ships, I think they could now rank as worthy members of your group.' (Memo from S.D. Cooke to MHT, 10/3/1950).

Vernon Industries were the British licensees of an American patent called 'Oxvar', a printed decorative finish which could be applied to any smooth surface, including hardboard, metal and glass. 'A condition of the licence is that the rollers must be made in the U.S.A. In fact, two rollers are used in the process: a plain one of resilient material which accepts the pattern from an ordinary steel roller and transfers it to the mastic material being laid up on the sheet..' (Notes of meeting between John Skelton, General Manager, and MHT, 29/3/1950).

Samples printed on hardboard and aluminium with a temporary hand-roller were submitted by J.W. Graham, Director, 28/4/1950 with a view to supplying panels for the Regatta Restaurant. However, Misha Black declined to incorporate them into the interior. The idea of using FPG patterns on trays, pelmets, tabletops, roof glass and television cabinets (for Ferranti) was floated, but probably never realised.

HM was later approached by Oxvar Ltd, the British subsidiary of the American firm who had developed the process. She entered into a contract with them on 4/10/1951 and supplied diagrams for the next three years, but none ever reached production. The contract ended on 31/3/1955.
Archives: AAD 1977/3/286-308; DCA 5384-1, 5384-2, 5384-3, 5384-4, 5396

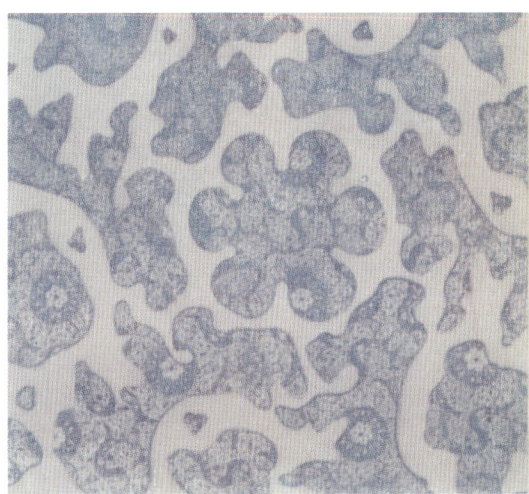

FPG 64
Insulin 8.27 'Oxvar' decorative finish
Roller-printed mastic applied to wood, plastics, metal or glass
Crystallographer: Dorothy Hodgkin
Designer: W.T. Higgins
Repeats: 17.8 x 10.2cm; 78.7 x 43.2cm (2 different scales); 1 colourway
Souvenir Book, p.4: 'By this offset process a decorated wearing surface can be applied to many different materials in furniture, wall-coverings and the like. The witty device of using the pattern twice, at two different scales, can be seen.'
Prototypes only. Considered for use on dumb waiters in Regatta Restaurant, but Warerite was selected in the end. Used on glass desk top in information bureau in Power and Production. Samples shown in FPG displays in the Regatta Restaurant, Dome of Discovery and Land Travelling Exhibition.

FPG 49 - 'Festival' plate, designed by Hazel Thumpston for R.H. & S.L. Plant

FPG 76 - Beryl 8.9 plate, designed by Peter Wall for Wedgwood

WARERITE

Tradename: Warerite
FPG code: A512

Manufacturers of laminated plastics based in Ware, Hertfordshire. Warerite Ltd. was a subsidiary of the American-owned company Bakelite Ltd., manufacturers of Bakelite phenolic resin plastics, invented by Leo Hendrik Baekeland in 1907, produced by the General Bakelite Co. from 1910. In 1939 the Bakelite Corporation was sold to the Union Carbide Corporation of America. Warerite specialised in high pressure laminated plastic sheets, in which offset litho-printed patterns were coated with durable thermosetting plastic resin (melamine). Warerite – like its competitor Formica – was extremely popular during the post-war period, widely used on tables, cabinets, counters and as panels on walls.

Warerite were invited to join the FPG on 13/12/1949 with a view to supplying furnishings and accessories for the Regatta Restaurant. Two patterns were approved for this purpose, advertised in *Design*, May-June 1951: 'Pattern Building from Molecular Structures. / Fresh impetus has been given to the development of new patterns for interior decorative schemes by the work of the Festival of Britain Pattern Group on crystallographic designs. As members of this Group we have developed several new patterns for the range of super-hard WARERITE Laminated Plastics. The two we illustrate [Hydrargillite and Haemoglobin] will be used for walls, doors and table tops in the Regatta Restaurant and other Festival buildings.' Warerite tops were also used on dumbwaiters made for the restaurant by Goodearl Brothers. Warerite was employed extensively at the Exhibition of Science for counters in the buffet, cloakrooms, information bureau and elsewhere. H.H. Lusty, Marketing Manager for Bakelite Ltd., specifically requested that Warerite should be credited rather than Bakelite.

Up to six patterns entered production in 1951 and at least one was sold to a furniture manufacturer. In 1955 there were plans to relaunch the collection, although whether this happened is unclear. All the patterns were designed by Martyn O. Rowlands (1923-2004), a pioneering industrial designer specialising in plastics who trained at the Central School of Arts and Crafts. In 1950 he was a designer at Bakelite, hence his involvement in the FPG. A Fellow of the Plastics Institute, he later worked independently and went on to win two Design Centre Awards (1958 and 1961). His most famous design was the Trimphone for Standard Telephones & Cables (1966).
Archives: AAD 1977/3/253-254, 331. DCA 5384-2, 5384-3

FPG 65, 66

FPG 65
Afwillite 8.45 decorative laminate
High pressure plastic laminate, printed under melamine
Crystallographer: Helen Megaw
Designer: Martyn O. Rowlands
Repeats: 5 x 7.8cm, 10 x 35.5cm (2 different scales)
Souvenir Book, p.6: 'the designer has superimposed the pattern again, to a larger scale.' *Architectural Review,* p.239;
Limited production in 1951.

FPG 66
Apophyllite 8.31 decorative laminate
High pressure plastic laminate, printed under melamine
Crystallographer: W.H. Taylor
Designer: Martyn O. Rowlands
Repeat: 5 x 5cm
Souvenir Book, p.11
Limited production in 1951.

FPG 67
Beryl 8.9 decorative laminate
High pressure plastic laminate, printed under melamine
Crystallographer: Lawrence Bragg
Designer: Martyn O. Rowlands
Souvenir Book, p.1; *Architectural Review,* p.239; *Homes and Gardens,* p.34 ; *Queen*
Limited production in 1951. Used as backplate on wall bracket lightfitting by GEC.

FPG 68

FPG 68
Haemoglobin 8.26 decorative laminate
High pressure plastic laminate, printed under melamine
Crystallographer: Max Perutz
Designer: Martyn O. Rowlands
Repeat: 2.8 x 2.8cm; 3 colourways
Collections: V&A Circ.45-1968; W.85-1975; Girton College, Cambridge
Souvenir Book, p.3; *Design* (advertisement)
In production in 1951. Twenty-four 7ft x 3ft sheets in Blackcurrant were specified for the Regatta Restaurant, 25/9/1950; used in coat recesses (14 panels) and on five service doors. Warerite gave HM a rounded triangular coffee table featuring this laminate: 'WX 158 Festival Pattern in blue. Table, pattern 4343 frame with top incorporating special panel designed by studio.' (Delivery note, 1/2/1952). 'I am writing to thank you for the coffee table, which arrived this week. It is a beautiful thing, and I am very proud to have it. It is a most useful size and shape too, just the right height when one is sitting in an armchair. My visitors have admired it very much.' (Letter from HM to H.H. Lusty, 10/2/1952).

FPG 69
Hydrargillite 8.33 decorative laminate
High pressure plastic laminate, printed under melamine
Crystallographer: Helen Megaw
Designer: Martyn O. Rowlands
Souvenir Book, p. 9; *Homes and Gardens*, p.35; *Design* (advertisement)
In production in 1951. This design was sold to a furniture manufacturer and used for tables. Circular table featured small hexagonal motifs arranged in a ring, forming one unit of a larger ring.

FPG 70
Insulin 8.27 decorative laminate
High pressure plastic laminate, printed under melamine
Crystallographer: Dorothy Hodgkin
Designer: Martyn O. Rowlands
1 colourway
Collections: V&A Circ.44-1968
Limited production in 1951. Shown in FPG display in Regatta Restaurant and possibly in Land Travelling Exhibition.

WARNER & SONS
Tradename: Warner Fabrics
FPG code: A505

Textile manufacturers based in Braintree, Essex. Warner & Sons were established in 1891, the successor to a series of silk-weaving companies run by Benjamin Warner. Originally based in Spitalfields, the company moved to Braintree in 1895, specialising in silk damasks and brocades. Cotton and wool furnishing fabrics were also produced; rayons were introduced during the 1920s; printed textiles expanded from 1927 onwards. Alec Hunter joined Warner as Production Manager in 1932, becoming a Director in 1943, but continued to design as well. Marianne Straub became a designer from 1950 after her company, Helios, was taken over by Warner. Renamed Warner Fabrics plc in 1987, it subsequently moved to Milton Keynes and was acquired by Walker Greenbank plc in 1994. The company archive was separated off in 2001 and sold to Braintree District Museum Trust in 2004; it now operates as the Warner Textile Archive. Warner Fabrics was sold to Turnell & Gigon in 2001; since 2006 the brand has been owned by Zimmer+Rohde.

Ernest W. Goodale, Director, was contacted about the FPG in October 1949 with a view to supplying furnishing fabrics for the Regatta Restaurant. He and Alec Hunter both attended the first FPG meeting, 16/12/1949. Marianne Straub attended an FPG meeting on 13/7/1950. Her design 'Surrey', based on Afwillite, was a key feature in the Regatta Restaurant. A nylon and rayon damask upholstery fabric called Welland, also designed by Straub, was intended for use on chairs in the restaurant. Although described in an advert as being based on a crystal structure, it does not match any FPG diagrams. The idea of coordinated FPG furnishing fabrics, carpets and wallpapers was floated by MHT, but Goodale said it would be problematic. 'Difficulties of marketing custom have frustrated attempts to match designs and colours between different manufacturers, but it remains a promising field for design development, especially if matching sets could be sold ready-made.' (*Souvenir Book*, p.6).

Although eye-catching, 'Surrey' was prohibitively expensive, so very little was sold. 'To be candid, we have found no interest at all on the buyers' part in the crystal structure idea and, although we have sold some of the materials we made for the Festival, they were sold either on their merits or because they were produced for the Festival.' (Letter from E.W. Goodale to MHT, 3/10/1951).
Archives: AAD 1977/3/227-231, 234; DCA 5384-2, 5384-3, 5384-4

FPG 71
Afwillite 8.44 furnishing fabric ('Surrey')
Jacquard-woven warp tapestry, made from wool, cotton and continuous filament rayon
Crystallographer: Helen Megaw
Designer: Marianne Straub
Pattern no: K 185/4; w.127cm; 4 colourways
Collections: V&A Circ. 306-1951+AB; V&A Circ.73-1968; WTA
Souvenir Book, p.6; *Design*, p.18 and p.32; *Homes and Gardens*, p.40; *Skinner's Record,* p.475

Limited production in 1951. Wholesale price 45s 6d per yard, plus 31s 9d purchase tax. Dark green and gold colourway used as curtains in the Regatta Restaurant (155 yards); also shown in FPG display. HM wanted to use this fabric for curtains at Girton College, but it was too expensive: 'Thank you for sending me the curtain patterns. The green one is lovely, and greatly admired, but alas! the price!... I am very grieved because it is a lovely thing.' (Letter from HM to E.W. Goodale, 20/5/1951.) HM later bought a 5yd length at a reduced price, 4/6/1951.

FPG 73
Haemoglobin 8.26 furnishing fabric ('Rings')
Screen-printed cotton
Crystallographer: Max Perutz
Pattern no: CM 52141; w.127cm; *repeat*: 35.5 x 66cm; 4 colourways
Collections: SM 1976-644/36
Souvenir Book, p.2: *Architectural Review*, p.238; *Homes and Gardens*, p.35
Limited production in 1951. Wholesale price 10s 6d, plus 6s 10d purchase tax. Used at Exhibition of Science for curtains in cinema foyer.

FPG 71

FPG 72
China Clay 8.2 furnishing fabric ('Harwell')
Jacquard-woven cotton and rayon
Crystallographer: G.W. Brindley
Designer: Alec Hunter
Pattern no: 774; *repeat*: 7.5 x 11.5cm; 4 colourways
Collections: V&A Circ.307-1951+AB; Circ.70-1968+A; WTA
Souvenir Book, p.12; *Science Guide*, p.40; *British Textiles*, p.51; *Skinner's Record*, p.475, 480
Limited production in 1951. Shown in Land Travelling Exhibition.

FPG 74
Nylon 8.54c furnishing fabric ('Helmsley')
Jacquard-woven cotton
Crystallographer: Charles William Bunn
Designer: Marianne Straub
w.127cm; *repeats*: 86.5 x 61cm, 43.2 x 30.5cm (2 different scales); 5 colourways
Collections: WAG T.10394; V&A Circ.308-1951+AB; WTA
Souvenir Book, p.12; *Science Guide*, p.40; *Skinner's Record*, p.475, 480
Limited production in 1951, probably increased later.

Insulin 8.27 diagram, crystallographer Dorothy Hodgkin

Josiah Wedgwood & Sons

FPG code: A525

Ceramics manufacturers based in Barlaston, Stoke-on-Trent. After five years in partnership with Thomas Whieldon, Josiah Wedgwood established his own pottery in Burslem in 1759. Capitalising on the success of his Queen's Ware, Jasperware and Black Basalt ranges, he teamed up with Liverpool merchant Thomas Bentley and built a large purpose-designed ceramics factory at Etruria, near Stoke-on-Trent in 1771. In 1940 the company relocated to Barlaston. From 1966 onwards Wedgwood acquired numerous other ceramics firms, diversifying into bone china and glass. In 1988 the Wedgwood Group merged with Waterford Crystal to become Waterford Wedgwood.

Wedgwood became involved in the FPG from November 1950 through their designer Peter Wall. As a final year student at the RCA in 1949-50, Wall had worked on a project to design crystal structure patterns for tableware under Professor R.W. Baker. Wall corresponded with HM at the time, and visited her in Cambridge. 'I shall be glad to see you to consult about crystal patterns. It is not a question of seeing these under a microscope; they are more comparable to a map. It would probably be most helpful if you brought ... your own sketches of the way you want to adapt them, so that we can see which of the adaptations retain their scientific meaning.' (Letter from HM to Peter Wall, 16/10/1949). On 13/4/1950, whilst on a placement at Wedgwood, he referred to ceramic lithographs featuring insulin and mica.

Wall joined Wedgwood later that year; they were subsequently invited to participate in the FPG and agreed to co-operate by producing his prototype designs. 'With the best will in the world, it would not be physically possible for us to finalise and produce a new lithograph pattern for dinnerware in time for the Festival, nor could we contemplate doing so unless we were satisfied that there would be an export demand. I understand from Baker, however, that it would be helpful if we could carry out a few sample pieces for lithographs being done at the Royal College by Peter Wall.' (Letter from Hon. Josiah Wedgwood to MHT, 30/11/1950). Victor Skellern, Art Director, later explained that they were severely constrained by government restrictions. 'As you know, we are forbidden to supply decorated goods to the Home Market, which used to be a very valuable testing ground, and all our efforts are devoted to filling overwhelming demands from the dollar markets for the established [ranges].' (Letter from Victor Skellern to MHT, 5/10/1951).

On 29/4/1951 HM wrote to Peter Wall about obtaining one of his designs: 'I should very much like to possess a copy of one of your plates – the one with a border made from the charts, in green – and I am writing to know whether you would either sell me the original, after it has been exhibited, or else whether I could get a copy made. If the copying is not too difficult, I would even like a set of them, and perhaps some to give as presents as well.' (Letter from HM to Peter Wall, 16/12/1950). He explained that the plate was a one-off but offered to present it to her as a gift, although it is unclear whether this ever happened. On 28/11/1951 Wall requested a diagram of the crystal structure of oxygen as he was working on some designs connected with Joseph Priestley. HM supplied a drawing and he sent her a sketch of his design.

Archives: AAD 1977/3/332-348; DCA 5384-2, 5384-3, 5384-4; 5396

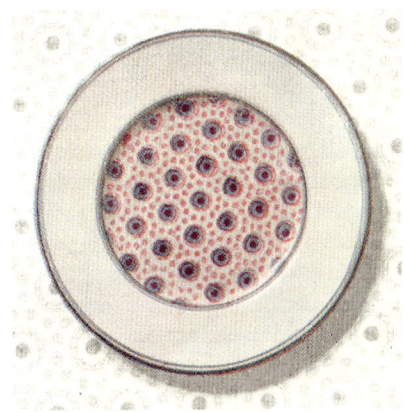

FPG 75
Haemoglobin 8.26 plate
Bone china, printed in overglaze enamels, gilded rim
Crystallographer: Max Perutz
Designer: Peter Wall
Repeat: 2.8 x 4.8cm
Souvenir Book, p.2; *Architectural Review,* p.238
Prototype only.

FPG 76
Beryl 8.9 plate
Bone china, printed in overglaze enamels
Crystallographer: Lawrence Bragg
Designer: Peter Wall
2 colourways
Collections: V&A Circ.41-1968 (lilac); Circ.42-1968 (grey)
Prototypes only. Shown in Land Travelling Exhibition. Although documented as Beryl 8.9, elements of this pattern bear a resemblance to Insulin 8.25 and Haemoglobin 8.26; it may be a free interpretation combining several crystal structures.

FPG 77 and FPG 79

FPG 77 Spinel 8.14 plate
Bone china, printed in overglaze enamels, gilded rim
Crystallographer: W.H. Bragg
Designer: Peter Wall
Homes and Gardens, p.34; *Queen*
Prototype only. The border features star-shaped motifs composed of criss-crossed strings of Spinel.

FPG 78
Spinel 8.14 tea cup and saucer
Bone china, printed in overglaze enamels, gilded rim
Crystallographer: W.H. Bragg
Designer: Peter Wall
Homes and Gardens, p.35
Prototype only. The border features zigzagging ribbons of Spinel.

FPG 79
Spinel 8.14 and Lithium Chlorate Trihydrate 8.15 tableware
Bone china, printed in overglaze enamels
Crystallographer: W.H. Bragg
Designer: Peter Wall
Souvenir Book, p.17; *Homes and Gardens,* p.34; *Queen*
These two crystal structures were juxtaposed on a series of prototypes. Various combinations of circular (Spinel) or diamond-shaped (Lithium Chlorate Trihydrate) motifs were used, in conjunction with diapered backgrounds (Spinel). Coffee pot, sugar bowl, plate, cup and saucer were shown in the FPG display in the Regatta Restaurant.

WOOD BROTHERS GLASS COMPANY
FPG code: A521

Glass manufacturers based in Yorkshire. Wood Brothers Glass Company was established at Worsborough, near Barnsley, around 1828 by a glassmaking family from the West Midlands. During the 19th century the factory produced lead crystal tableware, as well as bottles and other functional wares. By the mid 20th century they were focusing on scientific and medical glassware, glass tubing and perfume bottles.

Wood Brothers were invited to participate in the FPG at the suggestion of J.W. Chance because of their expertise in pressed glass, with a view to producing ashtrays for the Regatta Restaurant. They were approached on 28/4/1950 and initially prepared three designs, including Insulin. Delays in appointing the caterers delayed crucial design decisions, but eventually Pentaerythritol was chosen on 4/11/1950. A. Haslam Wood, Managing Director, generously offered to produce 700 ashtrays for free. Misha Black had misgivings about the design, but MHT persuaded him to agree and apologised for causing offence to the manufacturer. MHT was so pleased with the ashtrays that he sent one to his sister. Haslam Wood later asked him to intervene so that they could avoid paying purchase tax, as the ashtrays were supplied gratis. He remained extremely positive about the FPG: 'I feel that the way you have drawn manufacturers from various sections of British industry together on this design problem is admirable and I know many of us feel very happy about it and hope to continue our happy associations.' (Letter from A. Haslam Wood to MHT, 21 April 1951).
Archives: DCA 5384-2, 5384-3, 5384-4, 5396

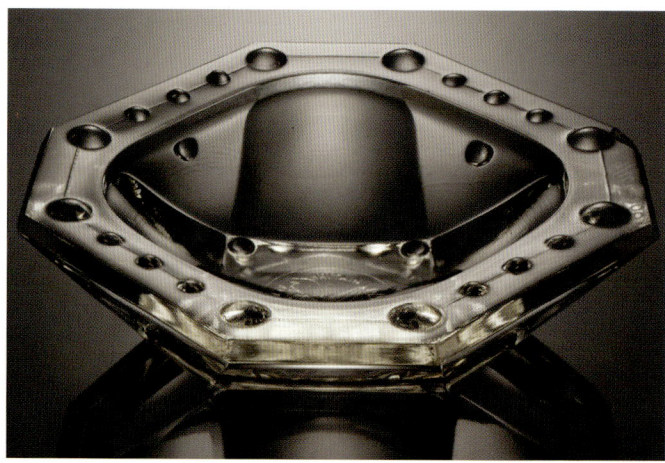

FPG 80
Pentaerythritol 8.18 ashtray
Press-moulded glass
Crystallographer: Gordon Cox
Designer: E. Sykes
d.12cm
Collections: SM 1993-421
Souvenir Book, p.10: 'This ashtray is the only product that attempts to portray the three-dimensional symmetry shown in a crystal structure diagram.'
700 ashtrays produced for Regatta Restaurant. Also featured in FPG displays in restaurant and Land Travelling Exhibition.

FPG 74 - Nylon 8.54c furnishing fabric ('Helmsley'), designed by Marianne Straub for Warner & Sons

APPENDIX
Other manufacturers and designers temporarily associated with the Festival Pattern Group

ALLEN SOLLY & CO.
Hosiery and knitwear manufacturers based in Nottingham. Joined FPG on 10/10/1950. Produced several knitted sock samples, including Lepidocrocite 8.52 and Quartz 8.10. Withdrew on 14/12/1950 because they felt they could not produce anything close enough to the original designs.

THE BRITISH ALUMINIUM COMPANY
Aluminium manufacturers, headquarters in London. Invited to participate on 12/6/1950, but said they could only produce simple embossed designs. Decided not to proceed due to manufacturing constraints.

THOMAS DE LA RUE & CO.
Tradename: Formica
Plastic laminate manufacturers, headquarters in London. Not invited to join, but authorised to produce designs based on crystal structures after April 1951 using FPG diagrams.

JOHN DICKINSON
Stationery manufacturers based in Hemel Hempstead. Invited to join but declined on 19/12/1950.

FURNITURE INDUSTRIES
Tradename: Ercol
Furniture manufacturers based in High Wycombe, Buckinghamshire. Lucien Ercolani, Director, attended first FPG meeting, 16/12/1949. Agreed to become involved and explored the idea of 'raised contour patterns'. After lengthy procrastination Ercol withdrew in October 1950.

HOGG & MITCHELL
Tradename: Old England
Clothing manufacturers, specialising in menswear and ties, based in Manchester. Hogg & Mitchell wrote to MHT on 28/6/1951 saying they would like to produce ties with crystal structure patterns in jacquard-woven crease-resistant rayon. On 16/7/1951 they selected two designs: Insulin 8.24 and Afwillite 8.43. They said the fabrics would be woven by one of their manufacturers and the ties would appear in 12 months' time.

LAWLEYS
Ceramics manufacturers based in Stoke-on-Trent. Invited to participate by MHT in March 1950, acting on a suggestion by Professor R.W. Baker. He later questioned Lawley's suitability and suggested that the COID should approach three ceramics manufacturers concurrently. The invitation to Lawley's was renewed on 23/10/1950, but they declined saying there was not enough time.

MICHAEL NAIRN & CO.
Linoleum manufacturers based in Kirkcaldy. Approached on 18/8/1949, but W.C. Gordon Black, Director, was not very positive; pulled out 29/11/1949.

GORDON RUSSELL
Furniture manufacturers based in Broadway, Worcestershire. Invited to participate on 6/12/1950 specifically to experiment with machine-routed decoration on veneers. Although they did not join, they were authorised to use FPG diagrams after 1/4/1951.

R.D. RUSSELL
Furniture and product designer based in London. Invited to experiment with crystal structure designs on furniture, 20/12/1950. Said he doubted whether the patterns were suitable for application on wood, but proposed trying them on a leather-topped library table.

WALKER & HALL
Manufacturers of cutlery and electroplate based in Sheffield. Invited to participate on 17/11/1949 with a view to supplying cutlery for the Regatta Restaurant. Colonel A.N. Lee, Managing Director, identified several crystal structure motifs for 'badging' on tableware and flatware. He and designer Victor Cowan attended the first FPG meeting, 16/12/1949. On 21/2/1950 he flagged up the need for a quick decision in order to produce the cutlery on time. On 29/3/1950 he complained about the amount of time and expense that the project was taking up and announced the firm's withdrawal.

JAS. WILLIAMSON & SON
Linoleum manufacturers based in Lancaster. Invited to participate on 13/12/1949 with a view to producing printed linoleum for the Regatta Restaurant. On 6/7/1950 Designs Manager, G. Sutton, sent some sample designs. Resigned from FPG in November 1950 in protest at the lack of suitable outlets for their products within Regatta Restaurant and Exhibition of Science.

FPG 59 China Clay 8.6 tie, designed by George Reynolds for Vanners & Fennell

FPG 36 - Boric Acid 8.34 wallpaper, designed by William Odell for John Line & Sons

CHAPTER FOUR

'PATTERN IN CRYSTALLOGRAPHY' by Helen Megaw, November 1946

(Edited transcript of an unpublished essay commissioned by the Design Research Unit)

We shall make pictures, when we have the skill,
Of the clear crystals that these rocks distill,
And draw fair patterns to enrich our night
With the inexorable curves of light.

We shall weave traceries as fine as lace
Of the minute events of time in space...

(Winifred Holtby)

Most people have at one time or another noticed and admired the 'frost flowers' traced on window-pane or pavement by the freezing of thin films of moisture or water-vapour. It is easy to understand that these patterns owe their beauty to the underlying regularity of the crystal structure. The symmetry of the structure, and the variety of patterns to which it can give rise, is shown when the ice crystals have a chance to grow undisturbed as they fall through cold air, growing into the beautiful six-rayed snowflakes so familiar in illustrations, though rarely seen in our damp climate. If the ice crystals grow under less uniform conditions, spreading out from the coldest part of the window-pane, the shapes that are formed are due to the combined effect on the growth of the regularity of crystal structure and the variability of local conditions.

Perhaps ice is the only familiar everyday example of these crystal growths; but the metallurgist meets them constantly when he examines under his high-power microscope the polished sections of the alloys in which he is interested... Here crystals of one chemical composition, with their own characteristic structure and habit, start to grow first while their surroundings remain molten metal, which in its turn solidifies rather later; or there may be more than two steps, each contributing to the final result.

In reality, of course, crystals have a three-dimensional regularity of structure, not merely the two-dimensional regularity so far considered, [but] since, however, we are only concerned here with patterns which can be represented on a plane surface, we do not need to go beyond two dimensions. Many crystals, like ice, are so thin that their patterns are in effect two-dimensional; for others... we can cut a section through our material and consider only the outlines that are shown up on that.

These examples illustrate the general principle that the outward shapes of crystals may depend greatly on local variations in the environment where they grew, and the regularity of their form may thus be obscured. But the fact that such regularity shows through at all is an indication of some underlying regularity in the structure of the material, in the arrangement of the atoms of which it is built. Such a pattern of building units was guessed at by the older crystallographers, by Robert Hooke, Newton's contemporary, and by [René Just] Haüy in the eighteenth century, long before the discovery of X-rays allowed any possibility of its verification. This inner regularity is not confined to materials which can grow into recognisable crystals. It is possessed by almost all solids, including substances like chalk and jewellers' rouge, which look to the eye like amorphous powders. It is this pattern, this regularity in the arrangement of atoms, which the X-ray crystallographer sets out to investigate.

In his popular lectures, Sir William Bragg (one of the pioneers of X-ray crystallography, on whose work much of our present knowledge is based) used to introduce the idea of crystal structure by showing a picture of wallpaper, and analysing the construction of its pattern. It is an example which many crystallographers today find it helpful to follow when they are explaining the subject to beginners (though they may show a considerable range of taste in their selection of the particular wallpaper pattern to be used as an illustration!)

A crystal structure, like a wallpaper, consists of a unit of pattern which repeats itself indefinitely. To pick out this unit (called by crystallographers the unit cell) from the pattern, we start with any point and look for three others, exactly like the first in themselves and in their surroundings, and so placed that when we join them up they enclose a parallelogram. In special cases this parallelogram may be a diamond with all its sides equal (this occurs quite frequently), a rectangle, or even a square (commonly the case in bathroom linoleums). With crystals, which have a three-dimensional lattice, our unit is actually a box, either rectangular or oblique-sided. Whatever it is, we can construct our whole fabric by putting together an indefinite number of these units side by side, all facing the same way round. To describe our pattern to anyone else, we need only give a complete specification of one unit. For judging its aesthetic qualities, we may need to add several repetitions, but this is merely a matter of mechanical copying of the original unit. Thus our task in investigating a pattern is narrowed down to the determination of the size and contents of its unit cell.

In the crystal structure, the contents of the cell are chemical atoms – the atoms of which the material is built up. In outlining our unit cell, we may cheerfully split the atom if we wish, placing half of it in one cell and half of it in the next, since the dividing line between the cells is just as imaginary as the earth's equator. In a number of substances, some of the atoms are tightly bound to certain of their neighbours, forming groups known as molecules; but it is not necessary to take this into account when describing the pattern (though it is very important when we look beyond the pattern to the chemical and physical properties of the material). Simple structures have only one or two atoms in their unit cell, but more complicated structures have very large numbers. Perhaps the most complicated so far in any detail is haemoglobin... There is a corresponding variation in the size of the unit cell, ranging from about three ten-millionths of a millimetre for aluminium metal up to [text missing] for haemoglobin; and even larger cell sizes have been measured. One cannot really imagine these very small distances, but it may give some idea of the numbers involved to say that a grain of common salt, the size of a full stop, contains about a million unit cells along its diameter. Even the largest known cell is very far below the limits of the vision with the most powerful ordinary microscope.

Our knowledge of the structures has been derived by more indirect methods. Instead of light, we use X-rays. A beam of X-rays falls on the crystal, which, because of the regularity of its repeated pattern, diffracts it in a regular way, and the diffracted rays are recorded as black spots or lines on a photographic film. Because many of the crystal patterns are highly symmetrical, the recorded photographs are often very symmetrical also... From the position and intensity of the spots and lines on the photographs it is possible to calculate what the size and contents of the unit cell must be.

The mathematical treatment is often difficult, and not all structures are equally straightforward to determine, but the methods used are so well-tried that in general when a result is achieved we can feel confident in its truth. Perhaps some day we shall be able to confirm these indirect methods by direct photography with the electron microscope, which has already shown us objects much below the limits of the ordinary optical microscope, and is theoretically capable of dealing with things much smaller still. Recently such a photograph has been taken of a crystal with a very large unit cell actually showing (though not yet in any detail) the outline of the cell... The crystal is of interest in itself, being the substance that produces a certain virus disease in bean plants. One can just detect in the photograph the shape of the unit cells, forming a kind of chess-board pattern...

When the crystallographer has determined the size and contents of his unit cell, he generally draws some sort of working diagram for his own use and to describe the pattern to other people. He may mark the positions of his atoms by a dot at the centre of each; more commonly, he will use a little circle round the centre, perhaps varying its size according to whether the atoms are large or small, or even drawing their radii to scale, to touch each other as they do in the actual crystal. In his notes or lecture diagrams, where he is not restricted by the black and white of print, he will often use different colours for the different kinds of atoms; this is the same sort of convention as that used by the map-maker when he paints his seas blue and the British Empire red. He may join his atoms by spokes, to show where there are strong forces between them holding them together. He may draw two or three such different diagrams to show up different features of the structure; but however he draws it, it is the same structure, with the centres of the atoms in the same positions.

Alternatively, he may be interested not only in fixing the positions of the atom centres, but also in knowing how the negative electric charge (which does the scattering of the X-rays) falls off in going from the centre to the outside of the atom. He will show this by plotting a contour map, the atoms being shown as mountain peaks of electron density, with ridges between atoms that are strongly held together and valleys in the empty sites where there are no atoms. Whether he puts in only a few contour lines or many, the position of the atomic centres will always remain unchanged. One might sum it up by saying that all diagrams of a given crystal structure, however different from one another they may appear at first sight, have the same kind of family resemblance as a series of maps of the British Isles designed to show separately such things as roads, railways, physical features, rainfall, and density of population...

It is noticeable, when one looks at a selection of patterns of crystal structure, that many of them are characterised by high symmetry. We say that a structure possesses symmetry if one part of the unit cell, when cut out, can be made to coincide with another part either by simply turning it round or else by reflecting it in a mirror. Repetition of a unit cell to form an extended pattern does not in itself introduce symmetry, but crystal structures and wallpapers do in fact generally show symmetry as well as repetition. For three-dimensional lattices it can be shown mathematically that there are 230 different combinations of the possible symmetry operations; for two-dimensional lattices, there are 19. This means that if we take a given unsymmetrical shape, say the letter J, and put it in turn into every possible type of plane unit cell, repeating it in each according to the appropriate set of symmetry operations, we shall get 19 different patterns. But in addition to that, we can in each case vary the size of the J and its exact location in the cell; or we can change the shape of our unsymmetrical object altogether, using a comma instead of a J, or a question-mark or a bunch of flowers or part of a scroll, or something still more elaborate. In this way we get some idea of the unlimited variety of possible patterns, while showing

that it is yet possible to clarify them into quite a small number of types according to their symmetry. The actual use which has been made of the different groups by designers was illustrated by Sir Lawrence Bragg in a lecture at the Royal Institution in 1940; he was able to find and display fabrics belonging to nearly all 19 groups. Some of the groups, of course, give rise to patterns which are more satisfying to the eye than others. In some the symmetry is too obvious to please, in others there is too little symmetry to give a balanced effect. On the whole one might say that the most satisfying groups are those which give a fairly even distribution of the elements of pattern over available ground – in other words, those which pack the shapes comfortably together. Now these are just the groups in which we are likely to find actual crystal structures, since it is one of the chief principles of the architecture of solids that the atoms or molecules of the material shall pack together without any great gaps or awkward misfits. (Exceptions only occur when there are particularly strong forces holding the atoms together in rings or baskets enclosing quite large holes.) Hence the crystal patterns actually observed are likely to include many which are pleasing to the eye.

The figures chosen to illustrate the variety of crystal patterns are all diagrams of actual structures, many taken from original papers, drawn in different ways to illustrate different features of the structure, but all showing the characteristic symmetry of their own group. A single unit cell would in each case have sufficed to describe the structure; but in many of the figures a large number of cells have been drawn to show the effect of repetition. The choice of these illustrations has been to some extent arbitrary; they are just a small selection from the vast amount of material already accumulated by crystallographers, which is being added to every day.

These examples are enough, I hope, to show that the crystallographer is now in a position to repay his debt to the wallpaper designer. If our path to the understanding of crystal structures has been made easier for many of us at its outset by the contemplation of wallpapers, we now, having explored further into the territory opened up, can bring back new material which we hope may be built up into fresh advances in fabric design. It is often put forward as a professed aim of science to gain control of the processes of nature by learning to understand their mechanisms; but to most scientists, perhaps, an appreciation, however inarticulate, of the pattern underlying these processes is the driving force of their work. For the crystallographer these patterns are readily translatable into visual terms. It is hoped that these few examples drawn from such a rich field may suggest to designers ways in which to broadcast to a wider public some of the aesthetic pleasure found in the subject by crystallographers themselves.

Archive sources: AAD 1977/3/36; DCA 5396, 9466

'Notes on Crystal Structure Diagrams' by Helen Megaw, 12 January 1950

(Transcript of a document circulated to members of the Festival Pattern Group)

All the diagrams under consideration have actually arisen in the course of scientific work, though the particular way of displaying them is such as to put more emphasis on their decorative aspects. The patterns all depend on the underlying regularity of the arrangement of atoms in crystalline solids.

Possible sources of patterns may be divided into three main groups:

(1) Photographs of actual crystals. The regularity of growth of crystals taking place in an irregular environment gives rise to very varied and beautiful designs, of which the most familiar example is frost on the window pane. Numerous photographs of surfaces of alloys are available, showing a great variety of effects.

(2) X-ray photographs produced by crystals. These are not shadowgraphs like the familiar pictures of a hand showing the bones. They occur when a narrow beam of X-rays falls on a small crystal (say about the size of a pin's head); the regular arrangement of the atoms splits the beam up into a pattern of diverging beams, which produce black spots or streaks on a photographic film. The particular pattern depends not only on the kind of crystal but on the experimental arrangement. Metal sheets containing large numbers of crystals have characteristic patterns also.

Both (1) and (2) offer a great variety of patterns, and search among the actual photographs available might give designs suitable for many purposes as separate motifs.

(3) Maps of crystal structures. This type of pattern offers more possibilities, and all the illustrations circulated belong to this group.

The crystal consists of an endless repetition of a moderately simple unit of pattern. As in a wallpaper, the units of pattern need not be isolated, but may join on continuously to one another. The unit of pattern which, by repetition, gives rise to the whole structure, may be picked out in various ways. The repetition is in three dimensions, but in a good many cases it is possible to show it two-dimensionally, either by picking out a single layer of atoms all at one height, or by projecting on to the paper all the atoms in one unit of pattern, and indicating their different heights (if desired) by some convenient convention.

Notice that the diagrams are of the nature of maps. We cannot see the atoms, but we can find out their positions in the unit of pattern and map them. Usual conventions are as follows:

(a) The centres of atoms are marked with small circles, and they are joined by lines indicating forces holding those atoms strongly to one another. Sometimes the circles may be omitted, leaving the junction of the lines to mark an atomic centre.

(b) Lines may be used to join up atoms of one kind grouped round an atom of another kind. The commonest groupings of this kind are tetrahedra, octahedra, or cubes. Again the atomic centres may be marked with circles, or they may be sufficiently indicated by the corners of the tetrahedra, etc.

(c) Larger circles may be drawn to show the amount of space taken up by atoms of each kind – these are commonly known as packing diagrams. They may or may not include joining lines.

(d) Lines (full or dotted) may be used to outline the pattern repeat unit.

(e) Atoms of different kinds are indicated by circles of different sizes, open or solid, or differently shaded. If colour is available it may be used to distinguish them, the choice of colour being quite arbitrary; distinctions of shading, size or kind or outline of circle may then, if desired, be dropped. Similarly bonds of different kinds may be distinguished by different kinds of joining lines, full or dotted, or coloured.

(f) Atoms at different heights may also be distinguished, if desired, by different shading or colour.

(g) Another method of showing heights uses V-shaped lines for the bonds. The thick end of the line is assumed to be the upper end in every case.

h) Contour maps are also possible, showing the actual distribution of matter in the atom, and not merely the position of its centre. A peak on such a map marks the centre of an atom. Maps may represent sections of a structure, or projections on the plane of the paper. In the latter case, if several atomic peaks overlap, they are not very easily distinguishable by eye, and appear as ridges or confused hilly regions. Sometimes contours below a certain chosen level are drawn dotted. These are known as 'electron-density maps'.

(i) Another kind of contour map is very important, but much less easy to interpret, or to visualise its meaning. It is a map of interatomic distances in the structure. It is of considerable importance scientifically, and is known as "Patterson map". These are often more complicated, and with less obvious peaks, than the electron-density map. Some of them make very satisfying designs.

It thus appears that the same structure can be represented in a very large variety of ways, just as a map of England may show roads, or rivers, or counties, or mountains or a combination of several of these at once. It is legitimate to show only those features of the map which one desires to emphasise for a given purpose; but it is not legitimate to change their positions, or to put in things which are not there, or to put in some things of one kind and leave out others exactly similar. Colouring may be done in any way, provided things which are identical are coloured identically.

A further choice concerns repetition of the pattern. It is legitimate to show one repeat unit, or one unit plus parts of its neighbours, or several units, or a whole extended area of repetitions. The limits can be drawn anywhere convenient, provided they enclose at least one repeat unit. Anything less than this becomes meaningless as a crystal structure pattern.

Archive sources: DCA 5384-1, 5395, 5396

CHAPTER FIVE

AN A-Z OF CRYSTAL STRUCTURES

- *All quotations are from Helen Megaw's notes and correspondence unless otherwise indicated.*
- *Diagram numbers are those used by HM; diagrams were numbered sequentially as they were adopted into the FPG scheme. All recorded diagrams are listed, but not all were used.*
- *No crystallographers were publicly identified at the time. Titles and institutions cited are those in 1951.*
- *'Common Knowledge': HM's term for crystal structures that had been widely published and interpreted by various crystallographers.*
- *Source: Primary published scientfic source for each diagram. References are in Bibliography A.*

ADENINE HYDROCHLORIDE

$C_5H_5N_5 \cdot HCl$

An organic molecule which plays an integral role in DNA; contributes to the double-helix structure by forming a bond with a molecule called Thymine on the other side of the DNA strand. Adenine is also a component of ATP (Adenosine triphosphate), a molecule of vital importance in cells because of its ability to phosphorylise (transfer phosphate to other molecules), thereby allowing energy to be released.

Crystallographer: Dr. June M. Broomhead

Laboratory of Chemical Crystallography, University of Oxford

8.47 Adenine hydrochloride

Ball-and-spoke structure
Source: Broomhead (1948), p.329, fig.8
Souvenir Book, p.0: 'Both diagrams above relate to the same material, as can be seen by comparing the unit of structure of the two patterns. The difference in character is due to the former [8.47] showing by dot and line the positions of the atoms and the forces binding them strongly together, while the latter [8.48] is a "Patterson map" showing by contours interatomic distances in the structure.'
Collections: AAD 1977/3/445-446 (drawings); WTA (dyeline)

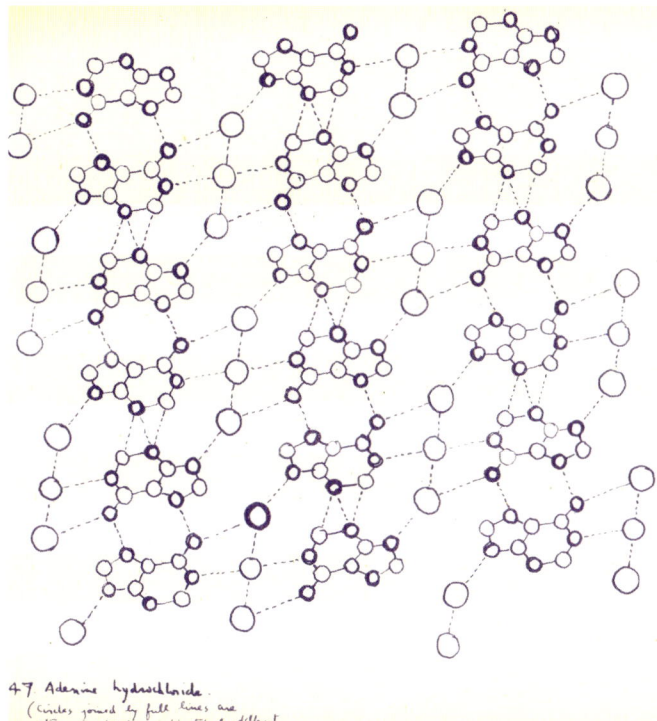

47 Adenine hydrochloride.
(Circles joined by full lines are atoms of 2 kinds, indicated by different ...)

8.47 Adenine hydrochloride diagram

8.48 Adenine hydrochloride

Patterson map
Source: Broomhead (1948), p.326, fig.3
Souvenir Book, p.0
Collections: AAD 1977/3/540, WTA (dyelines); AAD 1977/3/445 (drawing)
Diagram possibly used by R.H. & S.L. Plant

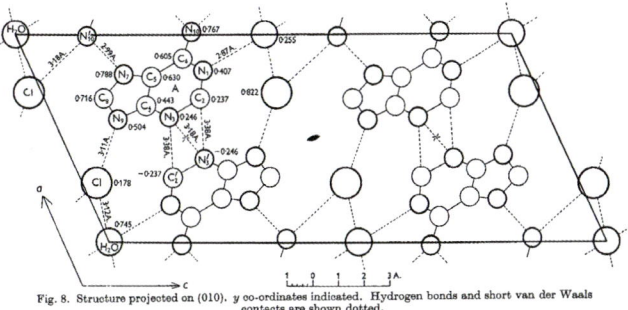

Fig. 8. Structure projected on (010). y co-ordinates indicated. Hydrogen bonds and short van der Waals contacts are shown dotted.

Source for Adenine hydrochloride 8.47
J.M. Broomhead, 'The Structure of Pyrimidines and Purines. II. A Determination of the Structure of Adenine Hydrochloride by X-ray methods', Acta Crystallographica, March 1948, vol.1, p.329, fig.8.
Copyright © International Union of Crystallography

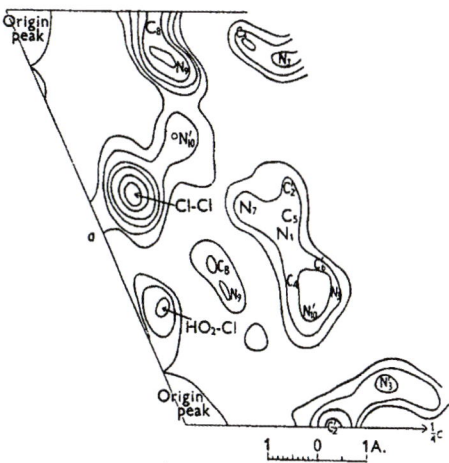

Fig. 3. Patterson projection on (010). Contours drawn at equal intervals on an arbitrary scale. The letters C and N indicate the positions of Cl–C and Cl–N vector peaks respectively.

Source for Adenine hydrochloride 8.48
J.M. Broomhead, 'The Structure of Pyrimidines and Purines. II. A Determination of the Structure of Adenine Hydrochloride by X-ray methods', Acta Crystallographica, March 1948, vol.1, p.326, p, fig. 3
Copyright © International Union of Crystallography

8.48 Adenine hydrochloride diagram

AFWILLITE

Ca$_3$Si$_2$O$_4$(OH)$_6$

Calcium hydroxide nesosilicate: a naturally occurring mineral characterised by glassy crystals. First identified at Dutoitspan Mine, Kimberley, South Africa in 1925; named after Alpheus Fuller Williams (1874-1953) of the De Beers Diamond Company. Afwillite can also be created artificially; it is one of the hydrated calcium silicates that form when cement sets to concrete.

Crystallographer: Dr. Helen Megaw

Cavendish Laboratory, University of Cambridge

8.37 Afwillite

Patterson map: '8.37 and 8.44 are contour maps of interatomic distances. 8.37 is a projection onto another plane at right angles to the first [8.44], related to it as elevations to plan.'
Source: 'The structure has not been published yet, apart from a preliminary note in Acta Crystallographica (1949). The full account is to be published in the same journal.' (AAD 1977/3/251)
Souvenir Book, p.1: 'has a niggling little Paul Klee flavour.'
Collections: WTA (dyeline)

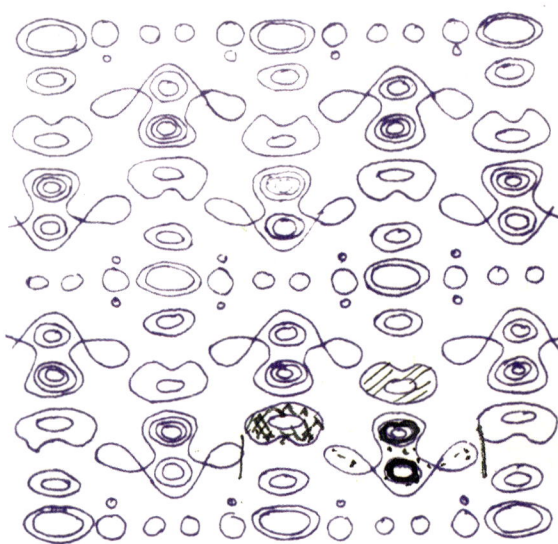

8.37 Afwillite diagram

8.38 Afwillite

Ball-and-spoke structure: 'Diagram 8.38 shows a projection of the structure. The small black circles represent silicon atoms, the larger black circles calcium atoms, and the other circles oxygen atoms, hydroxyl groups or water molecules.' (AAD 1977/3/251)
Related source: Megaw (July 1952), p.485, fig.6
Collections: AAD 1977/3/429 (coloured drawing); AAD 1977/3/515 (dyeline)

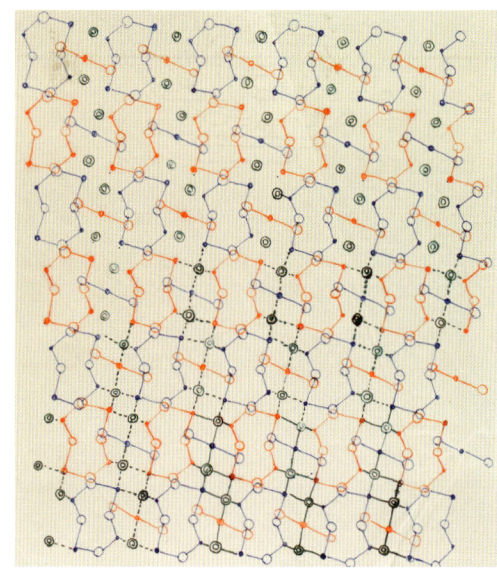

8.38 Afwillite diagram

8.42 Afwillite

Ball-and-spoke structure with tetrahedra
Collections: AAD 1977/3/509, WTA (dyelines)

8.43 Afwillite

Ball-and-spoke structure with tetrahedra: 'Atoms lie in 4 layers, equally spaced.'
Souvenir Book, p.1: [resembles] 'cocktail glasses and bubbles behind a bar.'
Collections: AAD 1977/3/514 (drawing); AAD 1977/3/510-511 (notes); WTA (dyeline)

Diagram used later by Hogg & Mitchell

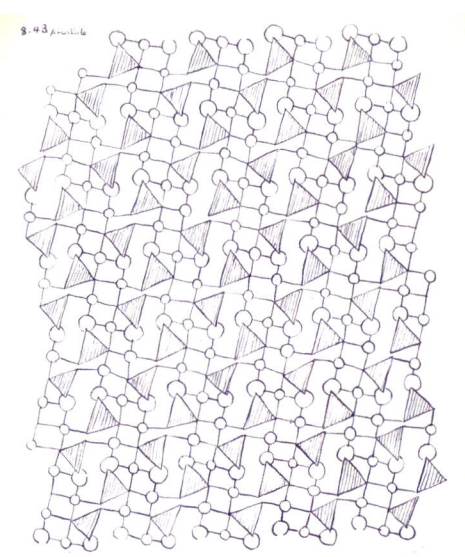

8.43 Afwillite diagram

8.44 Afwillite
Patterson map: 'map[s] of interatomic distances... a projection on the same plane as 37 and 45.'
Collections: WTA (dyeline)
Diagram used by Warner & Sons; and possibly by Arnold Lever

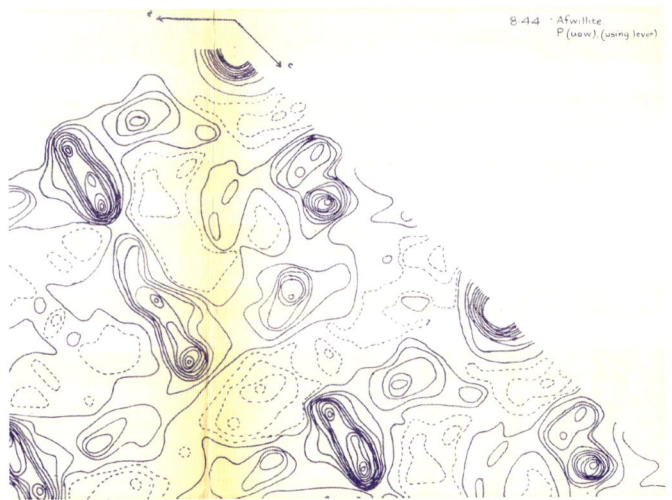

8.44 Afwillite diagram

8.45 Afwillite
Electron-density map, Fourier projection: 'The "peaks" with 6 or 7 contours are calcium atoms, those with 5 contours are silicon, and the others are hydroxyl and water.' (AAD 1977/3/251)
Related source: Megaw (July 1952), p.479, fig. 2
Souvenir Book, p.6; *Architectural Review*, cover, p.238
Collections: V&A Circ.78-1968 (dyeline)
Diagram used by British Celanese; John Line; Warerite

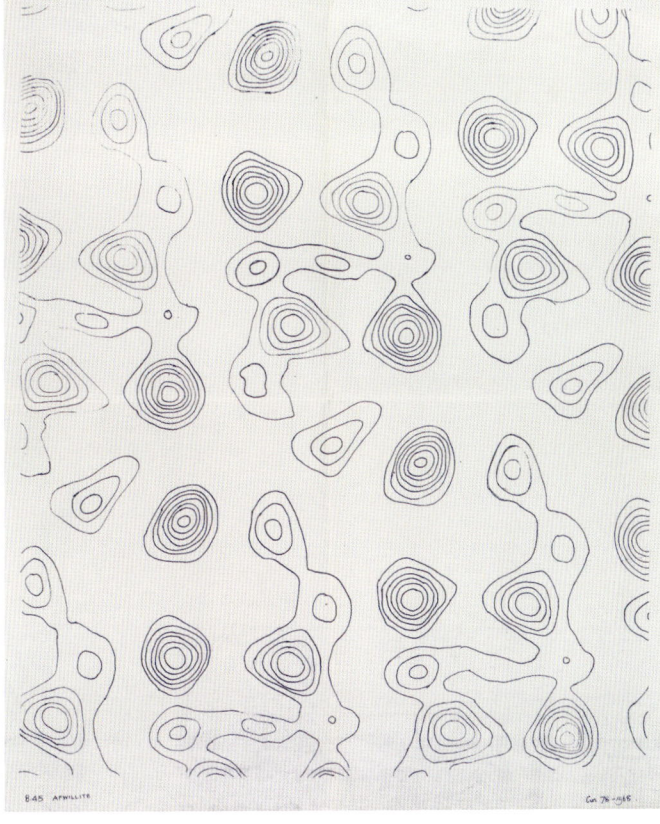

8.45 Afwillite diagram

ALUMINIUM HYDROXIDE (*see also* HYDRARGILLITE)
$Al(OH)_3$
Hydrated alumina: one of the most important - and most stable - compounds of aluminium ore (bauxite). Occurs naturally in the mineral gibbsite (also known as hydrargillite); also produced during the making of aluminium metal. Used in the production of various chemicals, including aluminium sulphate and aluminium fluoride; an ingredient in glass, glazes and fertilisers; a flame retardant filler in plastics; a bodying agent in paper, paint, inks and adhesives. Also has pharmacological applications: as an antacid (reducing acidity in the stomach); and as an adjuvant in some vaccines (improving antibody response).
Crystallographer: Dr. Helen Megaw
Cavendish Laboratory, University of Cambridge

8.4 Aluminium hydroxide
Ball-and-spoke structure
Sources: Bernal, Megaw (1935), p.400, fig.6; Bragg (1937), p.109, fig.64c
British Textiles, p.53
Collections: AAD 1977/3/487, 490 (annotated drawings); AAD 1977/3/571, V&A Circ.780-1968, T.446F-1977, WTA (dyelines)
Diagram used by Vanners & Fennell

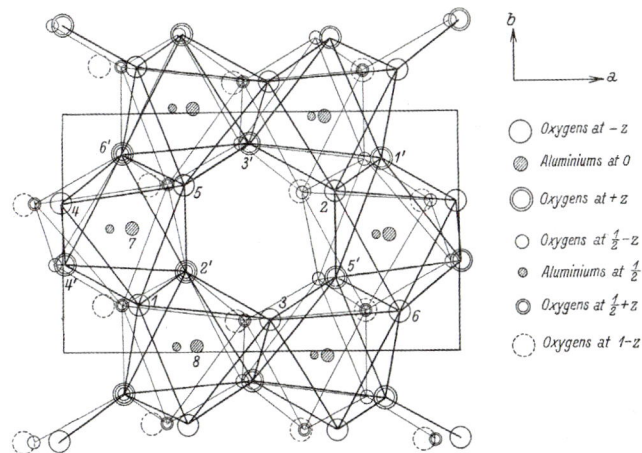

Fig. 6—Structure of $Al(OH)_3$ projected on (001), showing two layers of octahedra

Source for Aluminium hydroxide 8.4 + 8.8
J.D. Bernal, H.D. Megaw, 'The Function of Hydrogen in Intermolecular Forces', Proceedings of the Royal Society, Series A, 2 September 1935, vol.151, no.873, p.400, fig. 6

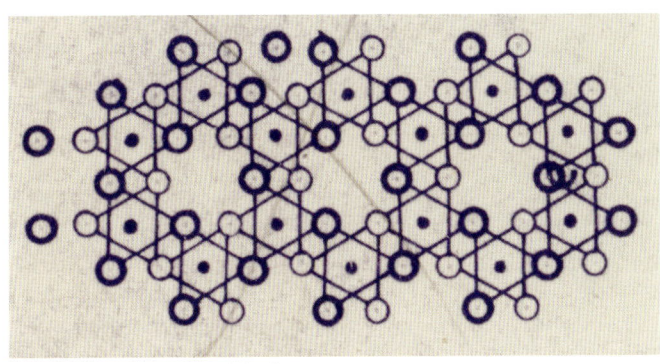

8.4 Aluminium hydroxide diagram

8.8 Aluminium hydroxide

Ball-and-spoke structure: 'Slightly different drawing of same structure as 4 and 5.' 'Originally coloured to distinguish the two overlapping sets of triangles, which are at different heights – all the large circles which are joined together being at the same height.'

Source: Bernal, Megaw (1935), p.400, fig.6; Bragg (1937), p.109, fig.64c
British Textiles, p.53
Collections: AAD 1977/3/572, WTA (dyelines)
Diagram used by E. Brain & Co.

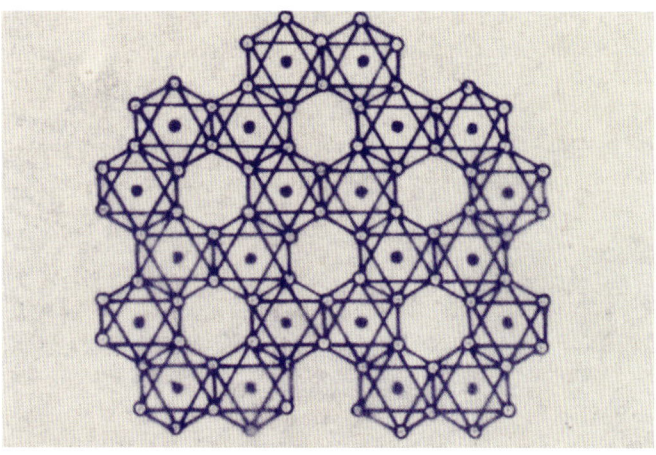

8.8 Aluminium hydroxide diagram

APOPHYLLITE

(K,Na)Ca$_4$Si$_8$O$_{20}$(F,OH)·8H$_2$O

Previously the name for one particular mineral, but redefined in 1978 to describe a group of phyllosilicates with similar chemical compositions. Prone to flake when heated; hence its name, from the Greek apophylliso, meaning 'it flakes off'. Composed of pyramid-shaped crystals which refract light, producing rainbow effects.

Crystallographer: Dr. W.H. Taylor
Cavendish Laboratory, University of Cambridge

8.30 [1-2] Apophyllite

Ball-and-spoke structure
Source: Bragg (1937), p.144, fig.83b; p.227, fig.125
Souvenir Book, p.10
Collections: AAD 1977/3/472 (notes); AAD 1977/3/474-477 (annotated drawings); V&A Circ.78J-1968, T.446F-1977, WTA (dyelines)
Diagram 8.30 [1] used by A.C. Gill; Chance Brothers

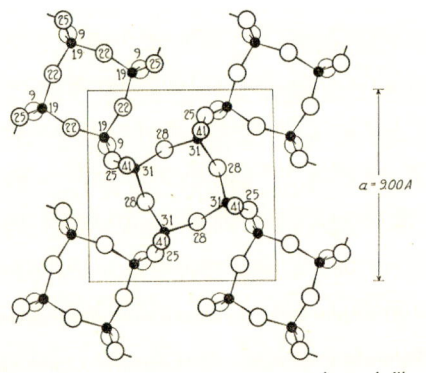

Source for Apophyllite 8.30
W.L. Bragg, Atomic Structure of Minerals, 1937, p.227, fig. 125

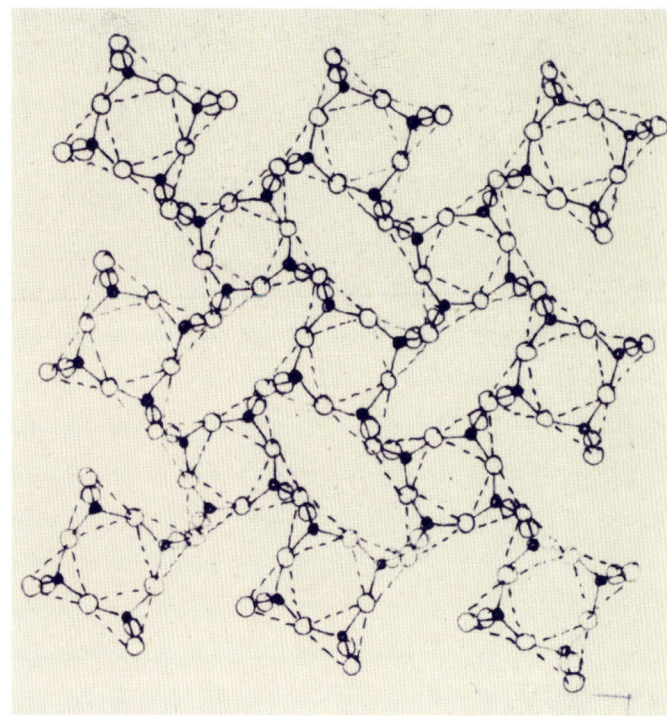

8.30 [1-2] Apophyllite diagram

8.31 [3-4] Apophyllite

Ball-and-spoke structure: 'This is an idealised version of 8.30. Take 4 regular tetrahedra and join into a ring.'
Souvenir Book, p.11
Collections: AAD 1977/3/478 (annotated drawing); AAD 1977/3/551, WTA (dyelines)
Diagram 8.31 [3] used by Warerite

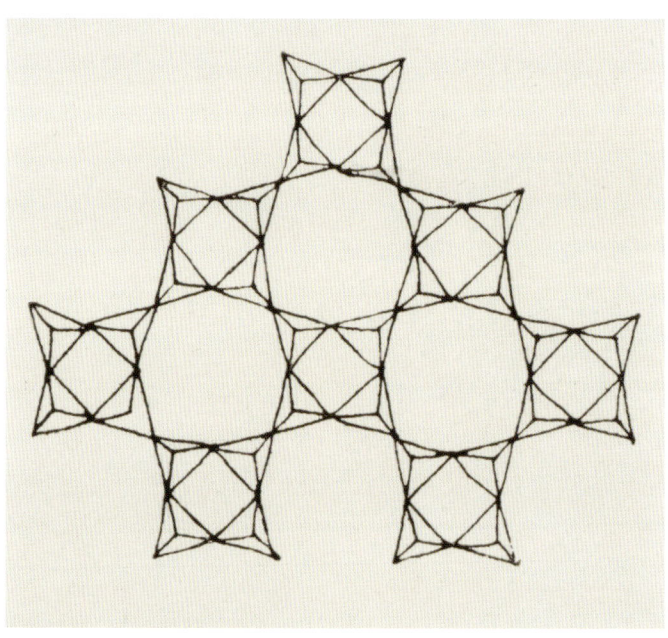

8.31 [3-4] Apophyllite diagram

BERYL

Be$_3$Al$_2$(SiO$_3$)$_6$

Beryl (beryllium aluminium cyclosilicate) is the generic name for a group of minerals composed of hexagonal crystals. Pure beryl (goshenite) is colourless; impurities produce coloured gemstones: aquamarine (blue); emerald (green); morganite (pink); bixbite (red); golden beryl (yellow).

Crystallographer: Professor Lawrence Bragg
Cavendish Laboratory, University of Cambridge

8.9 Beryl

Ball-and-spoke structure: 'Four different types of atoms, shown differently coloured in the original.'
Source: Bragg (1937), p.182, fig.104; Bragg (1962), p.141, fig.85
Souvenir Book, p.1; *Architectural Review*, p.238; *British Textiles*, p.53
Collections: AAD 1977/3/454 (drawing); AAD 1977/3/494-502 (annotated drawings); AAD 1977/3/518, V&A Circ.78N-1968, WTA (dyelines)
Diagram used by GEC; A.C. Gill; Warerite; Wedgwood

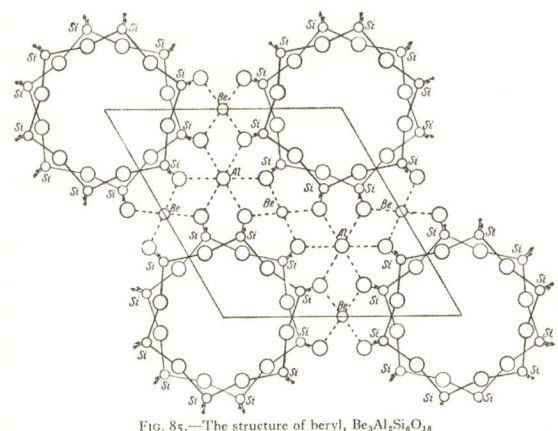

FIG. 85.—The structure of beryl, Be$_3$Al$_2$Si$_6$O$_{18}$

Source for Beryl 8.9
W.L. Bragg, The Crystalline State, 1962, p182, fig. 104

8.9 Beryl diagram

BISMUTH OXYCHLORIDE

BiOCl

A non-toxic heavy metal created as a by-product of lead and copper refining; white with an iridescent sheen. Widely used as a filler in cosmetics because of its binding and adhesive qualities and its pearlescent shimmer.

Crystallographer: Dr. Frederick Allan Bannister
Assistant Keeper of Minerals, British Museum (Natural History)

8.20 Bismuth oxychloride

Ball-and-spoke structure: 'Three kinds of atoms, each at a different height.'

BORIC ACID

B(OH)$_3$

A mild acid created when borate minerals (sassolite) are combined with sulphuric acid. Widely used as an antiseptic, flame retardant, wood preservative and insecticide. Also used in nuclear power stations to slow down the rate of fission in pressurised water reactors.

Crystallographers: Dr. Helen Megaw / Professor W.H. Zachariasen
Cavendish Laboratory, University of Cambridge / Department of Physics, University of Chicago

8.34 Boric acid

Ball-and-spoke structure with tetrahedra: 'All in one phase. Not suitable for 3-dimensional construction.'
Source: Bernal, Megaw (1935), p.405, fig.10
Souvenir Book, p.9
Collections: AAD 1977/3/460 (drawing); AAD 1977/3/485 (notes); V&A Circ.78B-1968, WTA (dyelines)
Diagram used by John Line

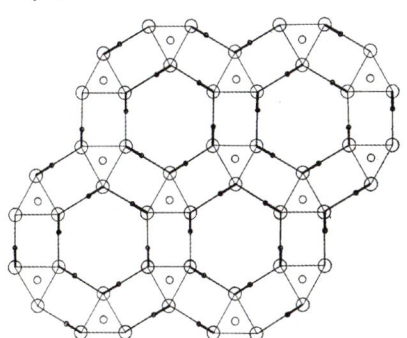

FIG. 10—Structure of one layer of B(OH)$_3$ parallel to (001), showing hydroxyl bonds.
○ boron; ○ oxygen; ○ hydrogen

Source for Boric acid 8.34
J.D. Bernal, H.D. Megaw, The Function of Hydrogen in Intermolecular Forces', Proceedings of the Royal Society, Series A, 2 September 1935, vol.151, no.873, p.405, fig. 10

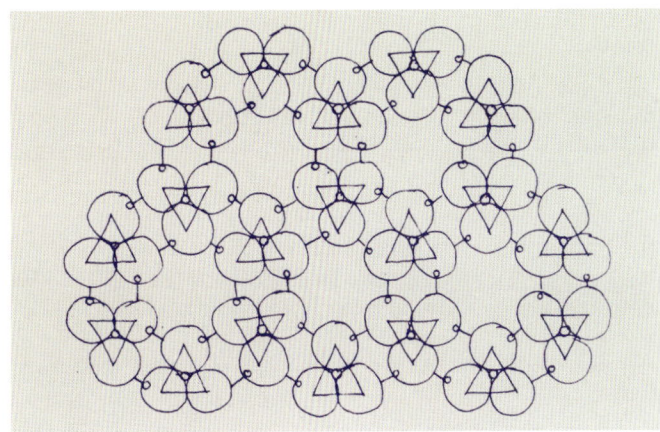

8.34 Boric acid diagram

BRUCITE

Mg(OH)$_2$

Crystalline mineral, also known as magnesium hydroxide, derived from magnesium metal. Closely related to aluminium hydroxide, brucite is white or green in colour and is a source of magnesia. Named after American mineralogist Archibald Bruce (1777-1818) who discovered it in 1824.

Crystallographer: Dr. Helen Megaw

Cavendish Laboratory, University of Cambridge

8.32 Brucite

Ball-and-spoke structure

Sources: Bragg (1937), p.108, fig.63; Bernal, Megaw (1935), p.397, fig. 4

Collections: AAD 1977/3/485 (notes); AAD 1977/3/487 (annotated drawings); AAD 1977/3/550, V&A T.446F-1977, WTA (dyelines)

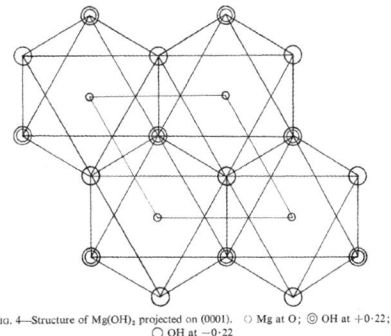

Fig. 4—Structure of Mg(OH)$_2$ projected on (0001). ○ Mg at O; ◎ OH at +0·22; ○ OH at −0·22

Source for Brucite 8.32

J.D. Bernal, H.D. Megaw, 'The Function of Hydrogen in Intermolecular Forces', Proceedings of the Royal Society, Series A, 2 September 1935, vol.151, no.873, p.397, fig. 4

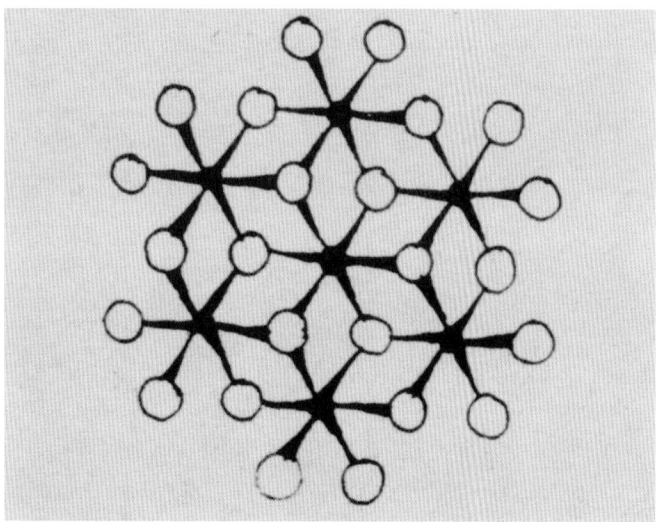

8.32 Brucite diagram

CELLULOSE

(C$_6$H$_{10}$O$_5$)$_n$ [general formula]

An organic compound composed of linked sugar molecules (polysaccharides), which provide the structural framework for plant cell walls. Because it is so fibrous, cellulose is difficult for humans to digest. A key component of wood and cotton, and a vital ingredient in textiles and paper-making, cellulose has been widely used in synthetic materials, from cellophane to viscose rayon to photographic film to nitrocellulose-coated fabrics.

Crystal structure: 'Common knowledge'

8.58 [a-f] Cellulose

Ball-and-spoke structures

Source: 'Rayon. I think this is essentially cellulose. The structure of this has been worked on by so many people that there are certain to be some usable diagrams.' (Letter from HM to MHT, 26/2/1950. AAD 1977/3/70)

Souvenir Book, p.0: 'The diagrams reveal that... cellulose is fibrous and stringy.'

Collections: AAD 977/3/533 (dyeline 8.58d-f)

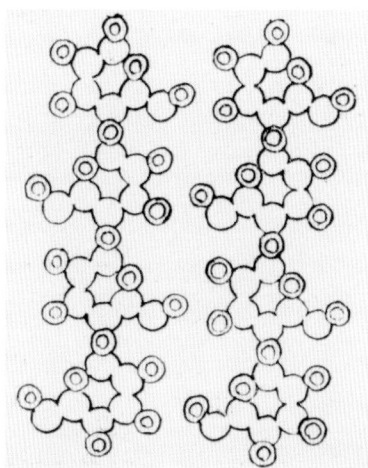

8.58 [a-f] Cellulose diagram

CHALK

CaCO$_3$

Soft, white, porous limestone formed from the mineral calcite, composed of either calcium carbonate or calcium sulphate. Traditionally used for writing on blackboards, marking lines on tennis courts and to aid grip in gymnastics.

Crystal structure: 'Common knowledge'

8.28 Chalk

Ball-and-spoke structure: 'Different outlines distinguish atoms at different heights.'

Source: Bragg (1937), p.119, fig.69

Collections: AAD 1977/3/554, V&A T.446F-1977, WTA (dyelines)

Diagram used by Vanners & Fennell

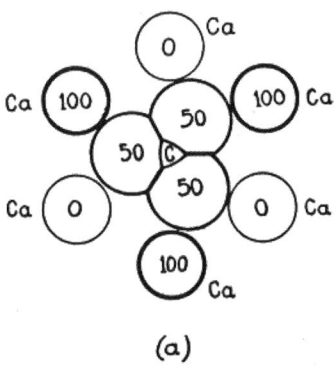

(a)

Source for Chalk 8.28

W.L. Bragg, Atomic Structure of Minerals, 1937, p.119, fig.69

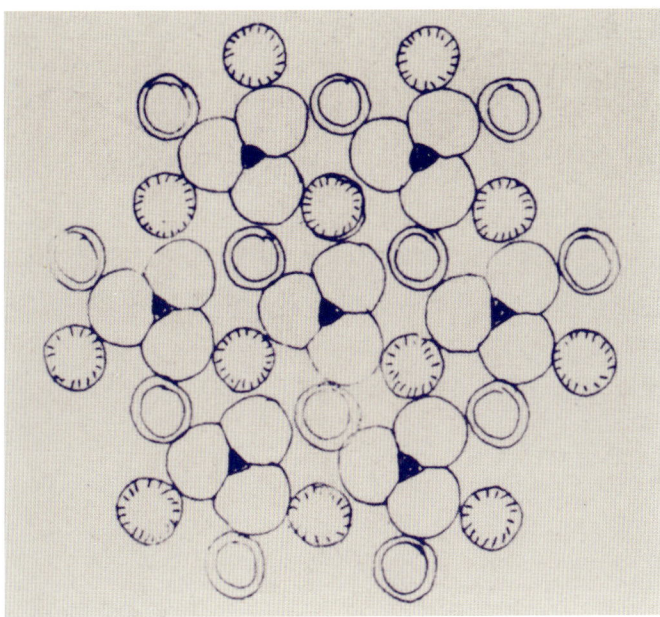

8.28 Chalk diagram

CHINA CLAY

$Al_2Si_2O_5(OH)_4$

A soft, white, powdery clay mineral, also known as kaolin or kaolinite, formed through the decomposition of aluminium silicates, especially felspar. A key ingredient in porcelain, china clay is widely used in pharmaceuticals and cosmetics, as a food additive and as a coating on paper.

Crystallographer: Dr. G.W. Brindley
Department of Physics, University of Leeds

8.2 China clay

Ball-and-spoke structure: 'These four, 2, 3, 6, and 7, are slightly different ways of displaying the same structure.'
Source: Brindley, Robinson (1946), p.249, fig.1: 'We now want permission to use some of your structures, in particular china clay. Actually it was in my original set of sample patterns, because I had been redrawing it myself in several different variants from your letter in Nature in 1945 or 6, so as to get a clear understanding of it, and having found how pretty it was I submitted it to show off the possibilities, and now they want to use it.' (Letter from HM to G.W. Brindley, 26/2/1950. AAD 1977/3/724)
Souvenir Book, p.12; *British Textiles*, p.53
Collections: AAD 1977/3/449, 524 (coloured drawings); AAD 1977/3/503-507 (annotated drawings); AAD 1977/3/571, V&A Circ.780-1968, WTA (dyelines)
Diagram used by Warner & Sons

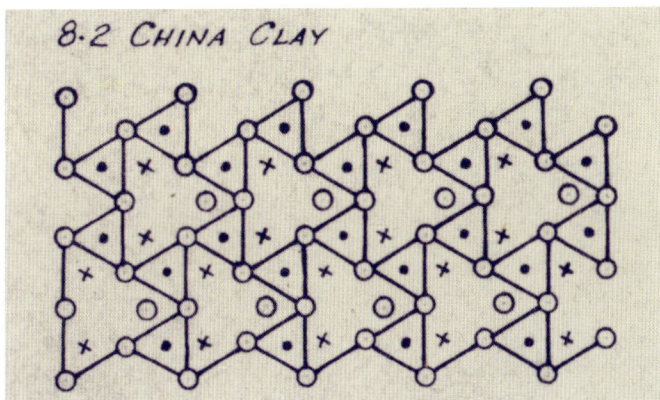

8.3 China clay

Ball-and-spoke structure: 'atomic positions'
Source: Brindley, Robinson (1946), p.249, fig.1
British Textiles, p.53
Collections: AAD 1977/3/524 (coloured drawing); 1977/3/571, V&A Circ.780-1968, WTA (dyelines)

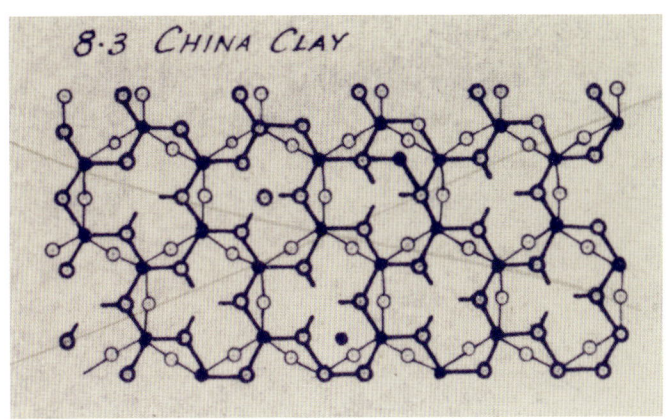

8.6 China clay

Ball-and-spoke structure: 'Hexagonal brucite in hydrargillite layer.' '[Vanners & Fennell] have another they call "china clay" – it's really the $Al(OH)_3$ layer from that [Clay Minerals 8.55] – but the slightly misleading title was my fault.' (Letter from HM to G.W. Brindley, 6/5/1951. AAD 1977/3/724).
Source: Brindley, Robinson (1946), p.249, fig.1: 'I have sketched in 2 unit cells of kaolinite (8.6) & you will see that a rather free use has been made of the scientific results.' (Letter from G.W. Brindley to HM, 19/2/1951. AAD 1977/3/726).
Souvenir Book, p.13; *British Textiles*, p.53
Collections: AAD 1977/3/572, V&A T.446F-1977, WTA (dyelines)
Diagram used by Dunlop; Old Bleach Linen Company; Stevens & Williams; Vanners & Fennell

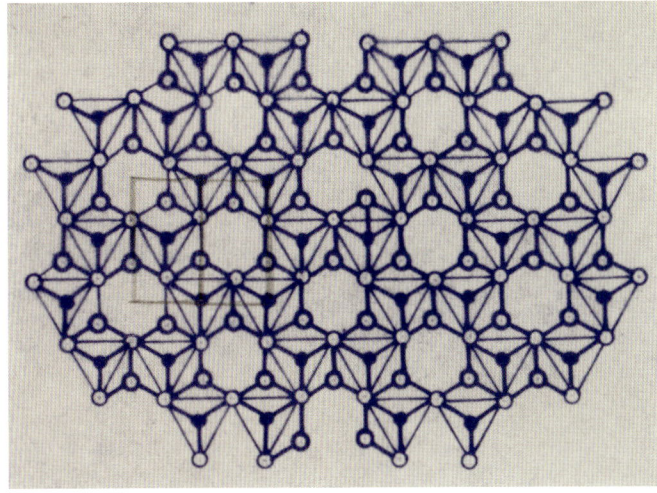

8.6 China clay diagram

8.7 China clay

Ball-and-spoke structure
Source: Brindley, Robinson (1946), p.249, fig.1
British Textiles, p.53
Collections: AAD 1977/3/572, V&A T.446F-1977, WTA (dyelines)

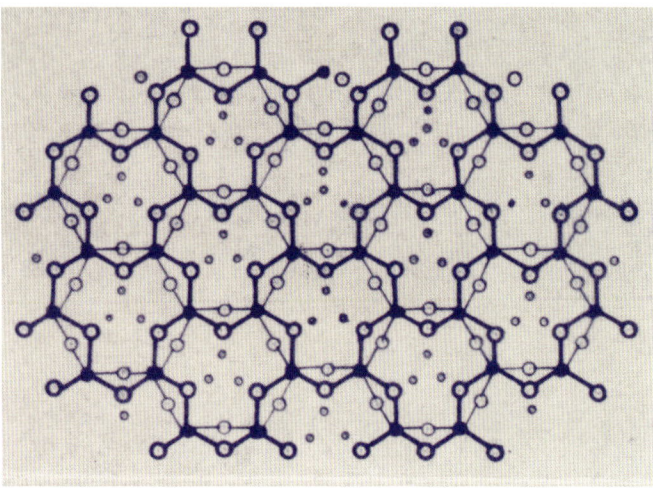

8.7 China clay diagram

CLAY MINERALS

$X_mY_4O_{10}(OH)_8$ [general formula, see below]

Fine-grained minerals, known as hydrous aluminium phyllosilicates, consisting of four main groups: kaolinite, smectite, illite and chlorite. Usually formed through the weathering of silicate-bearing rocks and the long-term effects of low concentrations of acidic solvents.

Crystallographer: Dr. G.W. Brindley
Department of Physics, University of Leeds

8.55 Clay minerals

Ball-and-spoke structure: '[Vanners & Fennell] have used your diagram which they call "clay minerals" but they have displaced the 2 layers 1⁄2 instead of 1⁄3 relative to each other – but it's very pretty.' (Letter from HM to G.W. Brindley, 6/5/1951. AAD 1977/3/724).
Source: Brindley et al (1950), p.410, fig.2: '$XmY_4O_{10}(OH)_8$, where *m* lies between 4 and 6, *X* and *Y* stand for positive ions in octahedral and tetrahedral positions respectively; usually X represents Mg with substitution by some Al, Fe, Cr or Mn, and Y represents Si and Al.'
British Textiles, April 1951, p.52
Collections: AAD 1977/3/526, V&A T.446F-1977, WTA (dyelines)
Diagram used by Vanners & Fennell

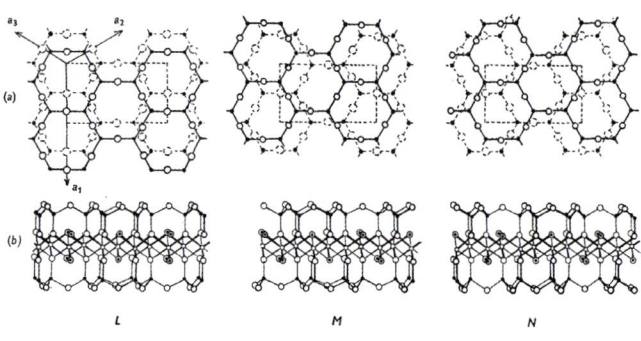

Source for Clay minerals 8.55
G.W. Brindley, B.M. Oughton, K. Robinson, 'Polymorphism of the Chlorites. I. Ordered Structures', Acta Crystallographica, November 1950, vol.3, p.410, fig.2
Copyright © International Union of Crystallography

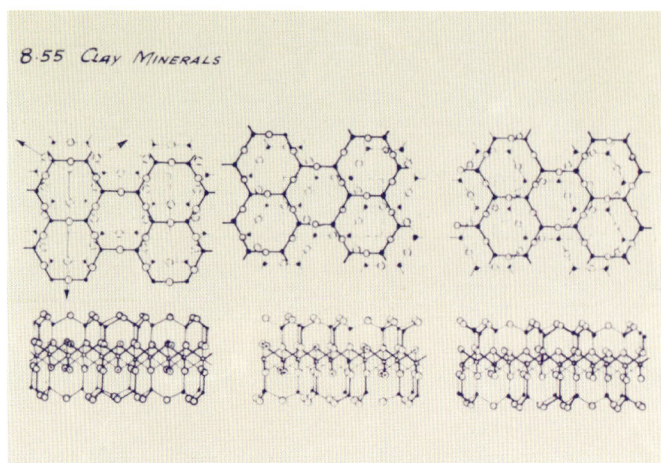

8.55 Clay minerals diagram

COPPER-ALUMINIUM ALLOY

$CuAl_2$

Aluminium to which copper has been added; a common alloy often used in products deployed at high temperatures, where it is crucial that the metal should not become elongated.
Crystal structure: 'Common knowledge'

8.19 Copper-aluminium alloy

Ball-and-spoke structure: 'Section of structure – atoms all of same kind at same height.'

CRISTOBALITE

SiO_2

A white or colourless crystalline mineral found in volcanic rocks, cristobalite has the same chemical composition as quartz (silicon dioxide), but a different structure. Cristobalite is only stable at very high temperatures. As it cools, its structure alters and, depending on how low the temperature drops, it is transformed into either tridymite or quartz.
Crystal structure: 'Common knowledge'

8.53 Cristobalite

Ball-and-spoke structure
Source: Bragg (1937), p.90, fig.52
Souvenir Book, p.8
Collections: AAD 1977/3/536, V&A Circ.78F-1968, WTA (dyelines)
Diagram used by G.A. Harvey

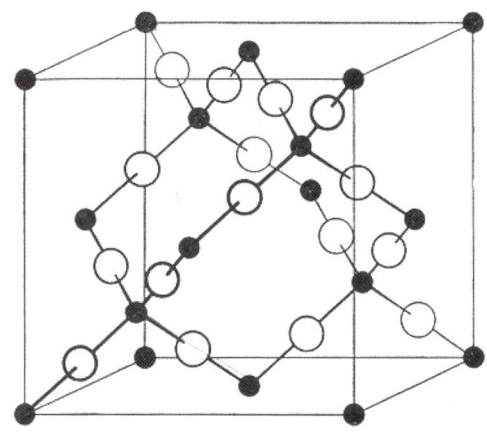

Source for Cristobalite 8.53
W.L. Bragg, Atomic Structure of Minerals, 1937, p.90, fig.52

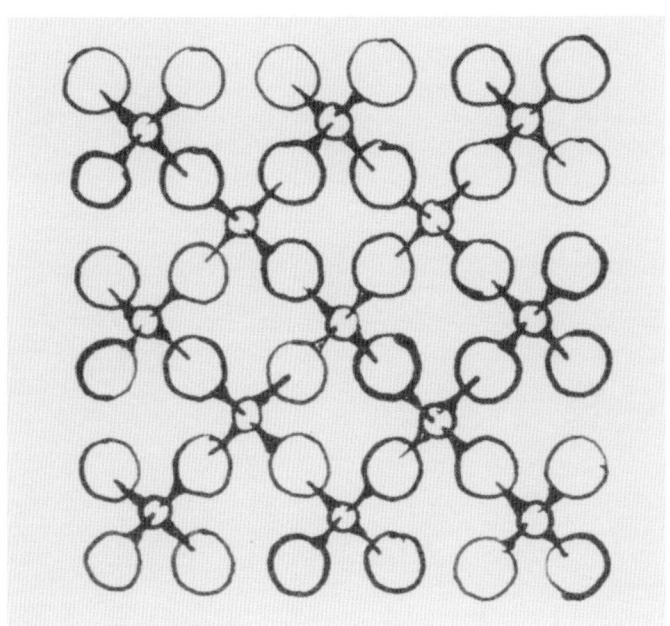

8.53 Cristobalite diagram

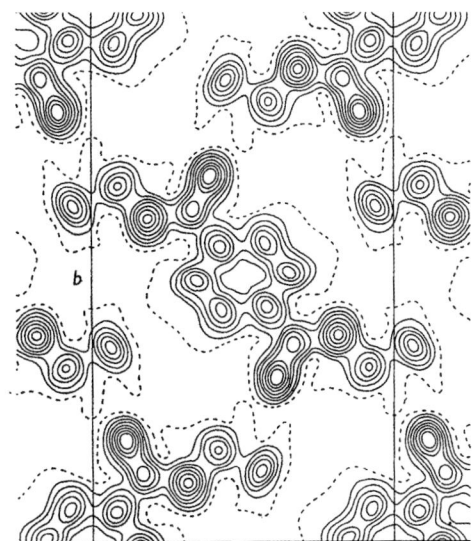

Source for Diethyl terephthalate 8.56
M. Bailey, 'The Crystal Structure of Diethyl Terephthalate', Acta Crystallographica, April 1949, vol.2, p.121, fig.1
Copyright © International Union of Crystallography

CYANURIC ACID

$(CNOH)_3$

A chemical, also known as triazine, created from the thermal decomposition of urea; commonly used in bleach, disinfectants and herbicides.

Crystal structure: 'Common knowledge'

8.12 Cyanuric acid

Ball-and-spoke structure: 'Three kinds of atoms.'

DIETHYL TEREPHTHALATE

$C_6H_4(COO \cdot C_2H_5)_2$

A precursor of polyester, the first synthetic fibre to be developed in Britain, produced by ICI from 1942 under the tradename Terylene. Diethyl terephthalate is a monomer or simple molecule, whereas polyethylene terephthalate (the thermoplastic resin from which polyester is made) is a polymer, a larger, more complex molecule.

Crystallographer: Myra Bailey

ICI Research Laboratories, Hexagon House, Blackley, Manchester

8.56 Diethyl terephthalate

Electron-density map, Fourier projection
Source: Bailey (1949), p.121, fig.1: 'Terylene. Preliminary structure work done on this only gives size of unit and does not allow construction of pattern. Would the F.O.B. Science Directorate be interested in diethyl terephthalate, which I think is the monomer which is polymerised to make terylene. This has quite a pretty structure, and though it was done in conjunction with I.C.I. it is by an author who may be able to give permission for its use without reference to higher authority.' (Letter from HM to MHT, 26/2/1950. AAD 1977/3/70)
Collections: AAD 1977/3/535, V&A Circ.78D-1968, WTA (dyelines)

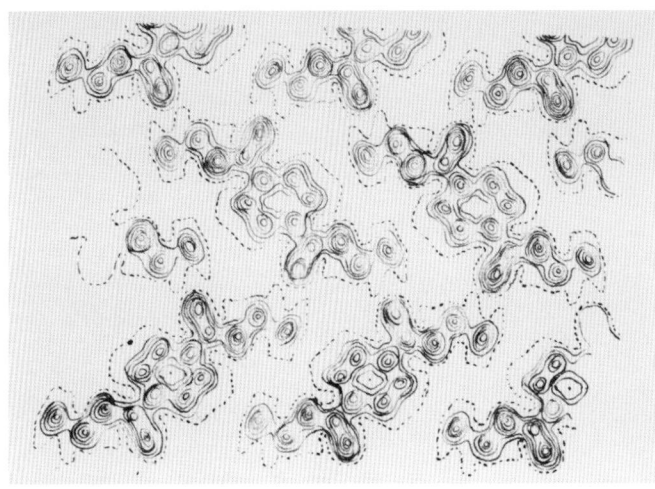

8.56 Diethyl terephthalate diagram

8.57 Diethyl terephthalate

Ball-and-spoke structure
Source: Bailey (1949), p.126, fig.5
Collections: AAD 1977/3/534, WTA (dyelines)

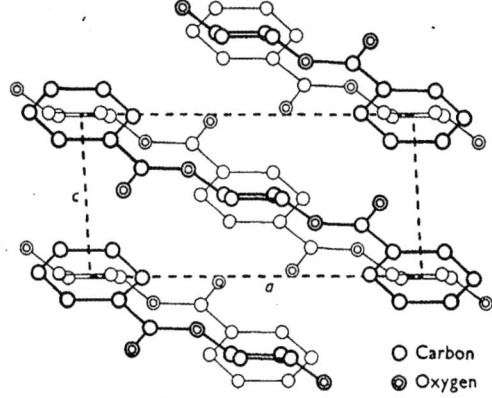

O Carbon
◎ Oxygen

Source for Diethyl terephthalate 8.57
M. Bailey, 'The Crystal Structure of Diethyl Terephthalate', Acta Crystallographica, April 1949, vol.2, p.126, fig.5
Copyright © International Union of Crystallography

DURANGITE

NaAl(A$_s$O$_4$)F

A translucent crystalline mineral, normally orange-red, consisting of a fluoarsenate of sodium and aluminium. Name after Durango, Mexico where it was discovered in 1869.

Crystal structure: 'Common knowledge'

8.11 Durangite

Ball-and-spoke structure
Collections: AAD 1977/3/451 (drawing)

GUANINE HYDROCHLORIDE

C$_5$H$_5$N$_5$O·HCI

An organic molecule known as a purine, which forms a vital constituent of DNA. Similar in structure to adenine hydrochloride, it codes genetic information in the polynucleotide chain.

Crystallographer: Dr. June M. Broomhead
Laboratory of Chemical Crystallography, University of Oxford

8.49 (1-3) Guanine hydrochloride

Probably ball-and-spoke structure: 'Lengthwise only.'
Collections: WTA (dyelines)

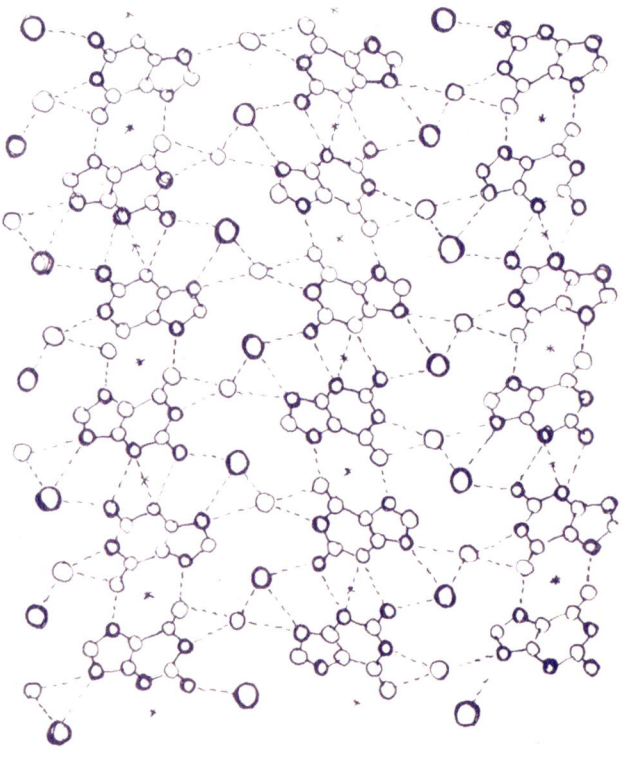

8.49 Guanine hydrochloride diagram

8.50 Guanine hydrochloride

Patterson map
Collections: AAD 1977/3/539, WTA (dyelines)

HAEMOGLOBIN

A complex protein which forms a major component of red blood cells; responsible for transporting oxygen from the lungs to other parts of the body. Haemoglobin molecules have four globular subunits, each containing a red pigment group, known as a haem, and a colourless protein residue, globin. At the centre of each haem is an iron atom which binds with oxygen. Each haemoglobin molecule can carry four oxygen molecules. When haemoglobin is oxygenated it is known as oxyhaemoglobin; after it has released oxygen to the tissues it is referred to as reduced haemoglobin or deoxyhaemoglobin.

Crystallographer: Dr. Max Perutz
Medical Research Council Unit for the Study of the Molecular Structure of Biological Systems, Cavendish Laboratory, University of Cambridge

8.26 Haemoglobin

Patterson map of reduced horse haemoglobin
Source: 'Patterson of reduced haemoglobin. Unpublished.' (Crystal Structure Diagram authorisation form, completed and signed by Max Perutz, March 1950. DCA 9466)
Possible related source: Bragg, Perutz (May 1952), p.325, fig.3b
Souvenir Book, p.3; *Architectural Review*, p.238; *British Textiles*, p.53; *Design*, Cover
Collections: AAD 1977/3/392, 517 (photographs); AAD 1977/3/463 (dyeline labelled 'Perutz')
Diagram used by Barlow & Jones; A.C. Gill; ICI Leathercloth; Arnold Lever; John Line; Spicers; Vanners & Fennell; Wedgwood; Warerite; Warner & Sons

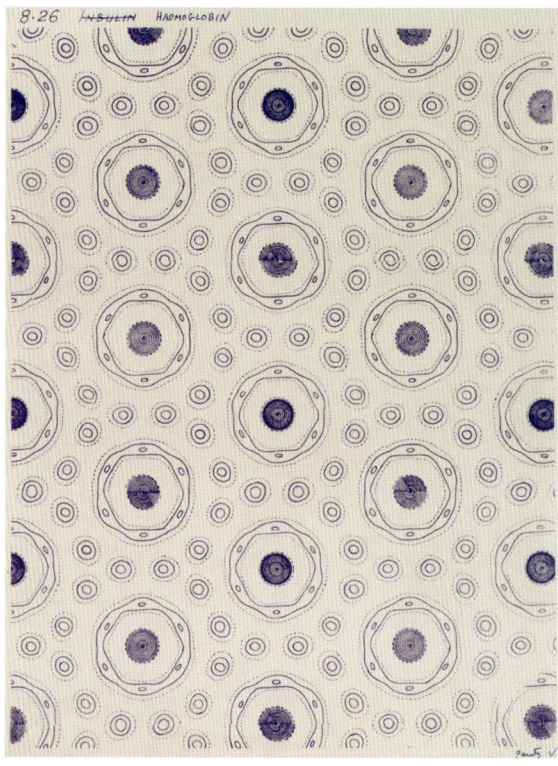

8.26 Haemoglobin diagram

HEXAMETHYLENE DIAMINE

$C_6H_{16}N_2$

A highly toxic and combustible chemical compound used as a raw material in the manufacture of nylon polymers and epoxy resins.

Crystallographer: Professor John Monteath Robertson

Department of Chemistry, University of Glasgow

8.40 Hexamethylene diamine dihydrochloride

Patterson map

Source: Binnie, Robertson (June 1949), p.181, fig.1

Souvenir Book, p.0: 'might almost be Tudor strap ornament'

Collections: WTA (dyeline)

Fig. 1. Patterson projection on (100).

Source for Hexamethylene diamine dihydrochloride 8.40

W.P. Binnie, J.M. Robertson, 'The Crystal Structure of Hexamethylenediamine Dihydrochloride', Acta Crystallographica, June 1949, vol.2, pp.180-188: p.181, fig.1

Copyright © International Union of Crystallography

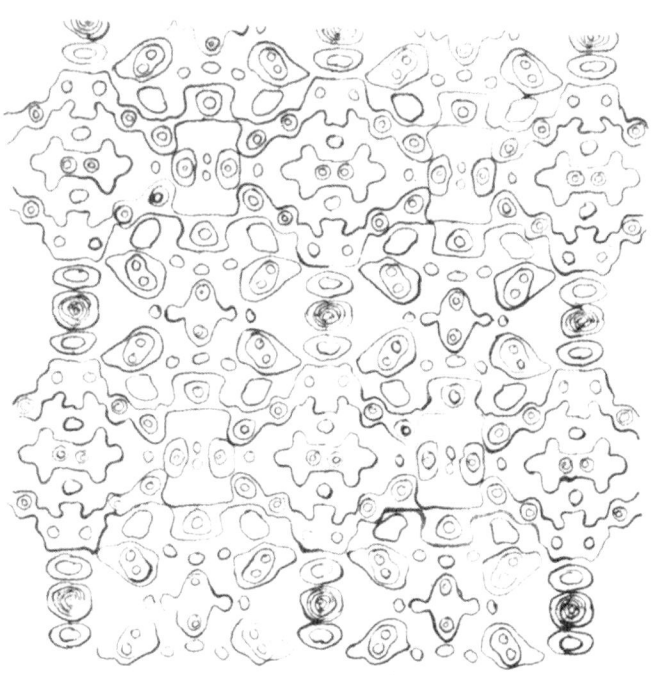

8.40 Hexamethylene diamine dihydrochloride diagram

8.41 Hexamethylene diamine dihydrobromide

Patterson map

Source: Binnie, Robertson (April 1949), p.117, fig.1

Collections: AAD 1977/3/547, WTA (dyelines)

Source for Hexamethylene diamine dihydrobromide 8.41

W.P. Binnie, J.M. Robertson, 'The Crystal Structure of Hexamethylenediamine and its Dihalides. Hexamethylenediamine Dihydrobromide', Acta Crystallographica, April 1949, vol.2, p.117, fig.1

Copyright © International Union of Crystallography

8.50 Hexamethylene diamine dibromide

Electron-density map

Source: Binnie, Robertson (April 1949), p.118, fig.3

Collections: AAD 1977/3/538, WTA (dyelines)

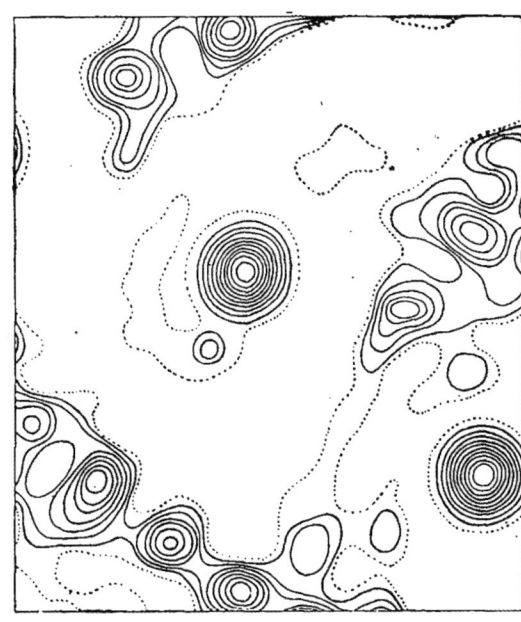

Source for Hexamethylene diamine dibromide 8.50

W.P. Binnie, J.M. Robertson, 'The Crystal Structure of Hexamethylenediamine and its Dihalides. Hexamethylenediamine Dihydrobromide', Acta Crystallographica, April 1949, vol.2, pp.116-120, p.118, fig.3

HORSE METHAEMOGLOBIN

Methaemoglobin is a variant of haemoglobin in which the iron atom is unable to carry oxygen. As a result the blood is brown rather than red.

Crystallographer: Dr. Max Perutz

Medical Research Council Unit for the Study of the Molecular Structure of Biological Systems, Cavendish Laboratory, University of Cambridge

8.23 Horse methaemoglobin

Patterson map, Fourier projection: 'Contour map of interatomic distances'

Source: Boyes-Watson et al., (1947), p.98, fig.3c

Souvenir Book, p.7

Collections: AAD 1977/3/415 (photograph); AAD 1977/3/555 (dyeline labelled 'Perutz'); WTA (dyeline)

Diagram used by British Celanese

Source for Horse methaemoglobin 8.23

J. Boyes-Watson, E. Davidson, M.F. Perutz, 'An X-ray study of horse methaemoglobin. I', Proceedings of the Royal Society, Series A, 26 September 1947, vol.191, no.1024, p.98, fig. 3c

8.23 Horse Methaemoglobin diagram

HYDRARGILLITE (*see also* ALUMINIUM HYDROXIDE)

$Al(OH)_3$

Aluminium hydroxide in crystalline mineral form; also known as gibbsite.

8.33 Hydrargillite

Ball-and-spoke structure

Sources: Bernal, Megaw (1935), p.400, fig.6; Bragg (1937), p.109, fig.64a

Collections: AAD 1977/3/485 (notes); AAD 1977/3/487, 490 (annotated drawings); AAD 1977/3/550, V&A T.446F-1977, WTA (dyelines)

Diagram used by A.C. Gill; Old Bleach Linen Company; Warerite; London Typographical Designers

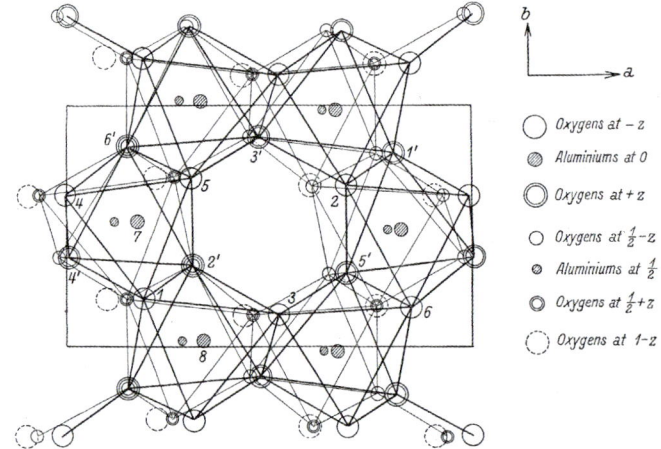

Source for Hydrargillite 8.33

J.D. Bernal, H.D. Megaw, 'The Function of Hydrogen in Intermolecular Forces', Proceedings of the Royal Society, Series A, 2 September 1935, vol.151, no.873, p.400, fig. 6

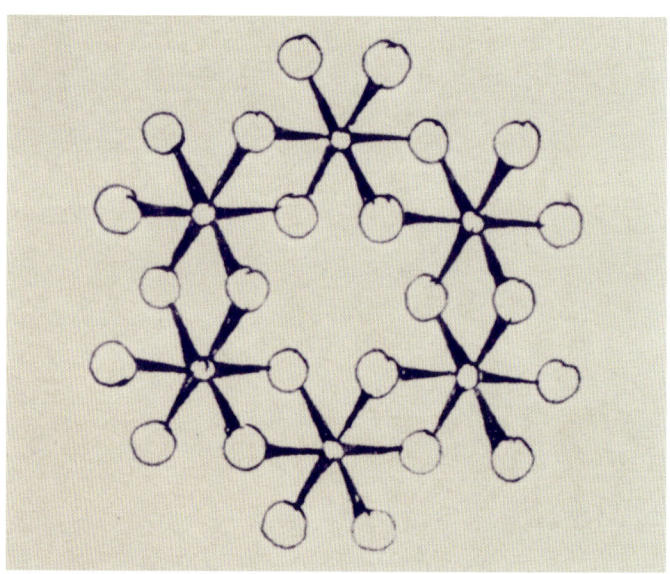

8.33 Hydrargillite diagram

INSULIN

A vital hormone secreted by the pancreas, activated by intake of food, insulin is a small but complex protein which monitors and controls the level of sugar in the blood. Insulin molecules circulate round the body where they are detected by insulin receptors in muscle, fat and liver cells, prompting them to absorb glucose from the blood and store it in the form of glycogen or fat. This is then used to make energy. Insulin deficiency results in diabetes mellitus, a condition where blood sugar rises to dangerously high levels.

Crystallographer: Dr. Dorothy Hodgkin
Laboratory of Chemical Crystallography, University of Oxford

8.24 Insulin

Patterson map: 'Contour map of interatomic distances'
Source: Dorothy Hodgkin: 'The particular patterns of yours which I want to ask for are the two of which I send you prints, and a third which is very like a refined version of the second.' (Letter from HM to Dorothy Hodgkin, 30/1/1950. AAD 1977/3/721)
Collections: AAD 1977/3/523 (dyeline labelled 'Dorothy'); WTA (dyeline)
Diagram used by ICI Leathercloth; and later by Hogg & Mitchell

8.24 Insulin diagram

8.25 Insulin

Patterson map: 'Contour map of interatomic distances'
Source: Crowfoot (1938), p.594, fig.1
Souvenir Book, p.5; *Architectural Review*, p.237; *British Textiles*, p.53; *The Times*, 24 April 1951
Collections: AAD 1977/3/390 (photograph); AAD 1977/3/471 (coloured dyeline labelled 'Dorothy'); V&A Circ.78L-1968, T.446F-1977, WTA (dyelines)
Diagram used by ICI Leathercloth; John Line; James Templeton; Vanners & Fennell

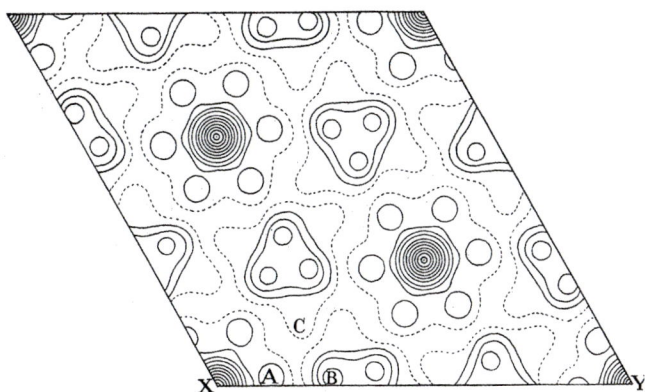

FIG. 1. *P(xy)* for insulin derived from Patterson-Fourier analysis. Contours at 5 units apart.

Source for Insulin 8.25
Dorothy Crowfoot, 'The crystal structure of insulin. I. The investigation of air-dried insulin crystals', Proceedings of the Royal Society, Series A, 18 February 1938, Vol.164, no.919, p.594, fig. 1

8.25 Insulin diagram

8.27 Insulin

Patterson map: 'Contour map of interatomic distances'

Source: Dorothy Hodgkin (see 8.24)

Souvenir Book, p.4; *Architectural Review*, April 1951, p.237; *Design*, Cover

Collections: AAD 1977/3/389 (photostat); AAD 1977/3/470 (coloured drawing); AAD 1977/3/512 (dyeline labelled 'Dorothy'); V&A Circ.78A-1968, WTA (dyelines)

Diagram used by Dunlop; A.C. Gill; John Line; Linoleum Manufacturing Company; Vernon Industries; Warerite

8.27 Insulin diagram

LEPIDOCROCITE

FeO·(OH)

Naturally occurring iron oxide-hydroxide mineral, also known as esmeraldite or hydrohematite, found in iron ore deposits exposed to weathering. Reddish-brown with a pronounced metallic lustre, lepidocrocite will rust under water due to its high iron content.

Crystallographer: Professor M. Stanislas Goldsztaub

Department of Mineralogy, University of Strasbourg

8.52 [a-c] Lepidocrocite

Ball-and-spoke structure: '3 different representations of structure'

Source: Bernal, Megaw (1935), p.404, fig.9; Bragg (1937), p.112, fig.65b

Collections: AAD 1977/3/457 (drawing); V&A Circ. 78G-1968, WTA (dyelines)

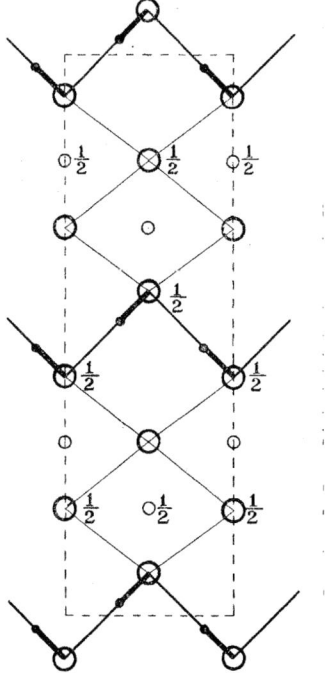

Source for Lepidocrocite 8.52

J.D. Bernal, H.D. Megaw, 'The Function of Hydrogen in Intermolecular Forces', Proceedings of the Royal Society, Series A, 2 September 1935, vol.151, no.873, p.404, fig. 9

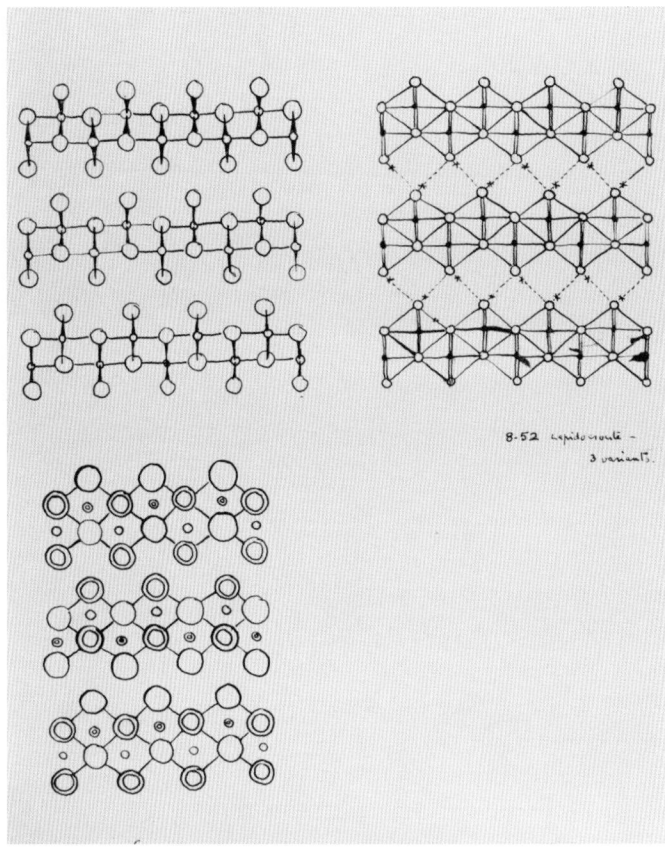

8.52 Lepidocrocite diagram

LITHIUM CHLORATE TRIHYDRATE

$LiClO_3\ 3H_2O$

An inorganic chemical compound with high water content, derived from the salts of chloric acid, containing an oxidised chlorine atom. Chlorates are active oxidisers but have limited applications because of their intrinsic instability.

Crystal structure: 'Common knowledge'

8.15 Lithium chlorate trihydrate

Ball-and-spoke structure with tetrahedra: 'Each diagram contains about four units of pattern, and could be extended or cut down. Diagrams differ only in style of drawing.'

Collections: DCA 9466 (photostat)

Diagram used by Wedgwood

8.15 Lithium chlorate trihydrate diagram

METALDEHYDE

$C_8H_{16}O_4$

An organic compound derived from acetic aldehye. A highly toxic chemical, metaldehyde is commonly used as a pesticide against slugs and snails.

Crystallographers: Professor Linus Pauling and J.H. Sturdivant

California Institute of Technology, Pasadena, California

8.22 Metaldehyde

Electron-density map

Diagram initially used by Dunlop, but later withdrawn: 'I have found the source of one of the original patterns you asked me to look up, because someone [Dunlop] had started to use it on, I think, a plastic curtain fabric. Unfortunately it comes from an American author, who can't easily be asked at this stage. I know him slightly and could ask him when the secrecy condition comes off, but wouldn't like to do it before that. The substance is metaldehye, and it was one of the very first batch.' (Letter from HM to MHT, 2/4/1950. DCA 5384-2)

MICA

$X_2Y_{4-6}Z_8O_{20}(OH,F)_4$ [general formula]; $KAl_1(AlSi_3O_{10})(OH)_2$ [muscovite]

A group of naturally occurring silicate minerals, which are chemically stable and heat-resistant, and which cleave easily due to the sheet-like arrangement of atoms. This diagram shows the structure of muscovite (white mica), the most common type of mica, widely used as an insulator in high-voltage electrical equipment and as an alterative to glass windows in stoves.

Crystal structure: W.W. Jackson / J. West

Physical Institute of the Victoria University of Manchester

8.16 Mica

Ball-and-spoke structure with tetrahedra: 'Atoms at corners of tetrahedra as well as centres of circles.'

Sources: Jackson, West (1930), p.211; Jackson, West (1933), p.160; Bragg (1937), p.207, figs.116-117

Collections: AAD 1977/3/563 (photostat)

Diagram used by Dobroyd

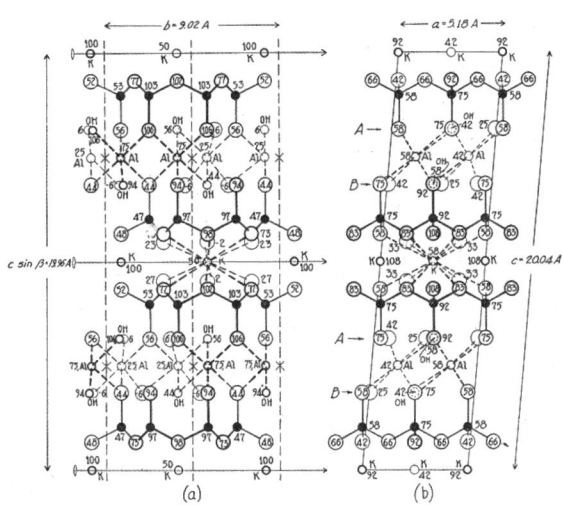

FIG. 117.—(a) The structure of mica (muscovite) projected at a plane normal to the *a* axis. The double sheets are seen, with K atoms between them surrounded by twelve oxygen atoms. (b) The same structure projected on (010). Note the staggering of the double layers which is responsible for the monoclinic angle of about 95°. Only half of the atoms are included.

In both figures a lower oxygen which would otherwise be eclipsed by an upper one is displaced slightly.

Source for Mica 8.16 and 8.35

W.L. Bragg, Atomic Structure of Minerals, 1937, p.207, fig.117

8.35 [1-3] Mica

Ball-and-spoke structure with tetrahedra: 'Atoms at corners of tetrahedra (appearing as triangles) as well as centres of circles.'

Sources: Jackson, West (1930), p.211; Jackson, West (1933), p.160; Bragg (1937), p.207, figs.116-117

Souvenir Book, p.8; *Architectural Review*, p.238, *British Textiles*, p.53

Collections: AAD 1977/3/430 (coloured drawing 8.35 [2]); AAD 1977/3/549, WTA (dyelines 8.35 [1-3])

Diagram 8.35 [2] used by Dobroyd; Dunlop; John Line

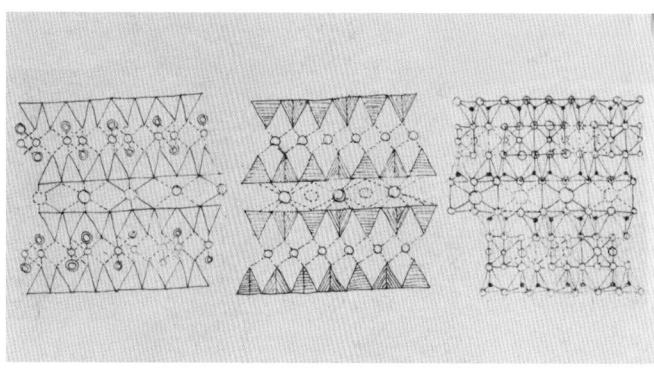

8.35 Mica diagram

MYOGLOBIN

An intricate globular protein which stores oxygen in muscle cells, responsible for the red colour in meat. Myoglobin molecules are about a quarter the size of haemoglobin and consist of a long chain of 153 amino acids and a haem group, containing iron atoms which bind themselves to oxygen molecules. High concentrations of myoglobin enable organisms to hold their breath for longer. Early research into the structure of myoglobin focused on whale and horse myoglobin, although the latter was later dropped. Whales need extra reserves of oxygen during their deep sea dives, so their muscle tissue is myoglobin-rich.

Crystallographer: Dr. John Kendrew

Medical Research Council Unit for the Study of the Molecular Structure of Biological Systems, Cavendish Laboratory, University of Cambridge

8.46 [a-b] Whale myoglobin

Patterson maps

Source: 'Proc. Roy. Soc. A 1950 (March) in part (some unpublished)' (John Kendrew, Crystal Structure Diagram authorisation form, March 1950. DCA 5395)

'Myoglobin is a red pigment found in muscle cells and, like haemoglobin, combines reversibly with oxygen. The myoglobins are much more difficult to crystallize than the haemoglobins, and though crystallization of several species has been reported in the literature, it is only recently that single crystals large enough for X-ray analysis have become available, derived from the horse and now from the whale.' Kendrew (1948), p.326

Collections: WTA (dyelines 8.46a-b)

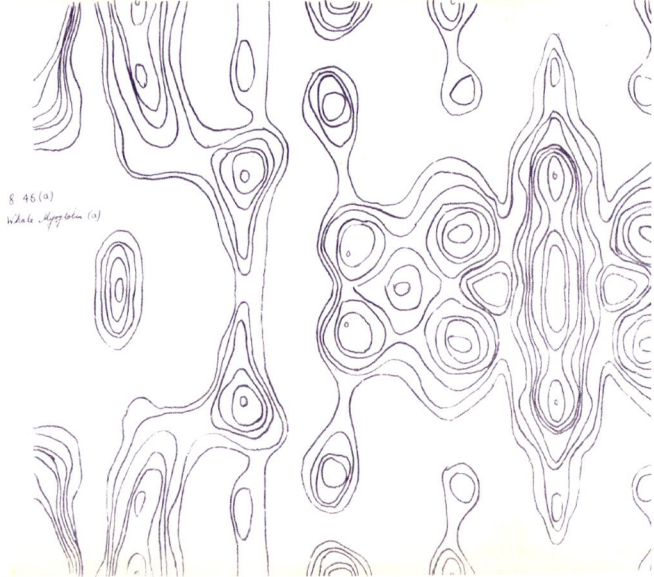

8.46 [a] Whale myoglobin diagram

8.46 [c-g] Horse myoglobin

Patterson maps (8.46c, d, f, g) and Fourier projection (8.46e)

Sources (8.46c): Kendrew (1950), p.74, fig. 7: Bragg et al (1950), p.350, fig.20a

Source (8.46d): Bragg et al (1950), p.350, fig.20b

Source (8.46e): Kendrew (1950), p.79, fig.10

Source (8.46f): Bragg et al (1950), p.353, fig.22a-b

Source (8.46g): Bragg et al (1950), p.353, fig.22c

Souvenir Book, p.7 (8.46c, g)

Collections: AAD 1977/3/441 (drawing 8.46f); AAD 1977/3/442 (drawing 8.46g); AAD 1977/3/443, 466 (drawings 8.46d); AAD 1977/3/444 (drawing 8.46e); AAD 1977/3/465 (drawing 8.46c); AAD 1977/3/541, V&A Circ.78I-1968, WTA (dyelines 8.46c); AAD 1977/3/542, WTA (dyelines 8.46d); AAD 1977/3/543, WTA (dyelines 8.46e); AAD 1977/3/544, WTA (dyelines 8.46f); AAD 1977/3/545, V&A Circ.78H-1968, WTA (dyelines 8.46g)

Diagrams (8.46c, f, g) used by ICI Leathercloth; possibly used later by A.C. Gill (AAD 1977/3/327)

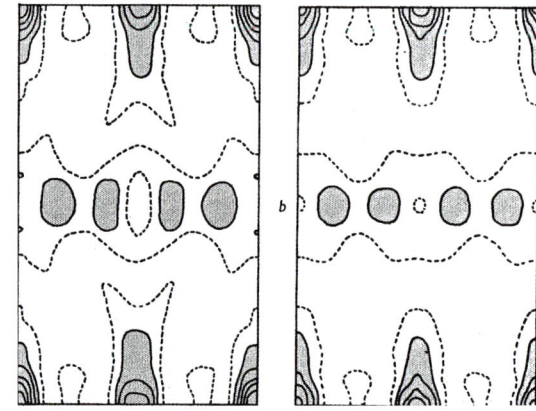

Source for Horse myoglobin 8.46c-d

W.L. Bragg, J.C. Kendrew, M.F. Perutz, *'Polypeptide Chain Configurations in Crystalline Proteins', Proceedings of the Royal Society, Series A, 10 October 1950, vol.203, no.1074, p.350, figs. 20a-b*

8.46 [c] Horse myoglobin diagram

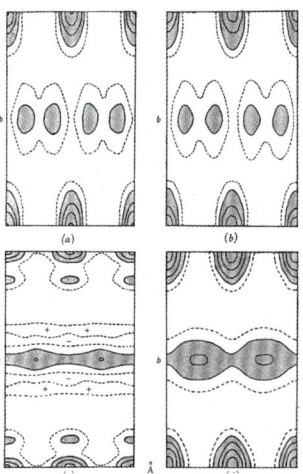

Source for Horse myoglobin 8.46f-g

W.L. Bragg, J.C. Kendrew, M.F. Perutz, *'Polypeptide Chain Configurations in Crystalline Proteins', Proceedings of the Royal Society, Series A, 10 October 1950, vol.203, no.1074, p.353, figs. 22a-c*

8.46 [g] Horse myoglobin diagram

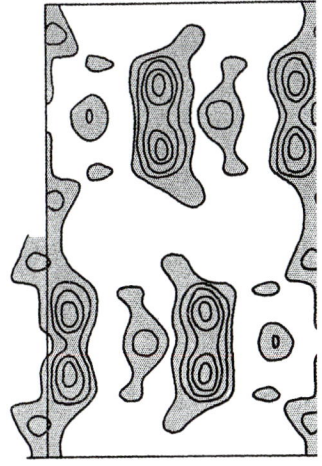

Source for Horse myoglobin 8.46e
J.C. Kendrew, 'The Crystal Structure of Horse Met-myoglobin. I. General Features: The Arrangement of the Polypeptide Chains', Proceedings of the Royal Society, Series A, 7 March 1950, vol. 201, no.1064, p.79, fig.10

NYLON

NHCO [general formula]; $NH(CH_2)_6NHCO(CH_2)_4CO$ [Nylon 66]
A family of long-chain thermoplastic polyamides invented by Wallace Carothers at Du Pont in 1935. The crystal structure in this diagram is Nylon 66 (polyhexamethylene adipamide), created by a condensation reaction between adipic acid and hexamethylene diamine. Nylon can be extruded, cast or injection-moulded, but its main use is in synthetic fibres, ranging from fine silk threads to tough carpet yarns. To produce fibres, nylon chips are melted and filtered, then extruded through a spinneret as filaments. When cold, the spun fibres are drawn out to the required thickness by twisting and stretching between rollers.
Crystallographer: Charles William Bunn
ICI (Alkali Division), Northwich; ICI (Plastics Division), Welwyn Garden City

8.54 [a-d] Nylon
Ball-and-spoke structures: 8.54b: '3 kinds of atoms shown by different symbols'; 8.54c: 'showing 3 layers'; 8.54d: 'showing 2 layers'
Source: Bunn, Garner (1947), p.51, fig.13: 'Nylon. Has been published... by a man from I.C.I., but is so simple that it can be reconstructed from the text plus general knowledge, without direct use of the diagram. Against that, it is rather dull and drab, but perhaps with careful attention to thickness of lines, sizes of circles, and such like factors it could be

made reasonably decorative.' (Letter from HM to MHT, 26/2/1950. AAD 1977/3/70)
Souvenir Book, p.12; *British Textiles*, p.52
Collections: AAD 1977/3/461-462 (drawings); AAD 1997/3/527, WTA (dyelines 8.54d); AAD 1977/3/530, V&A Circ.78E-1968, WTA (dyelines 8.54c); AAD 1977/3/531, WTA (dyelines 8.54b); AAD 1977/3/532, WTA (dyelines 8.54a)
Diagram 8.54c used by John Line; Warner & Sons

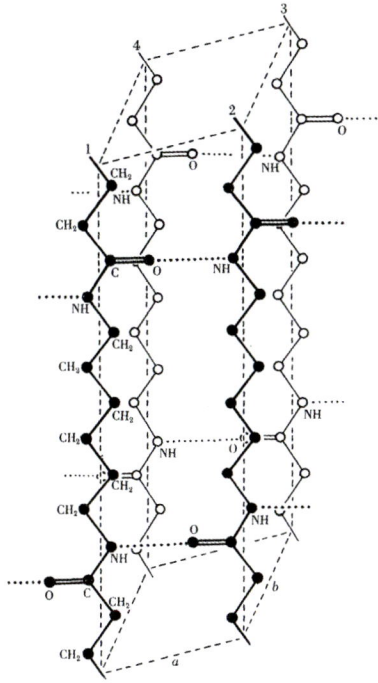

Source for Nylon 8.54
C.W. Bunn, E.V. Garner, 'The crystal structure of two polyamides ('nylons')', Proceedings of the Royal Society, Series A, 27 March 1947, vol.189, no.1016, p.51, fig.13

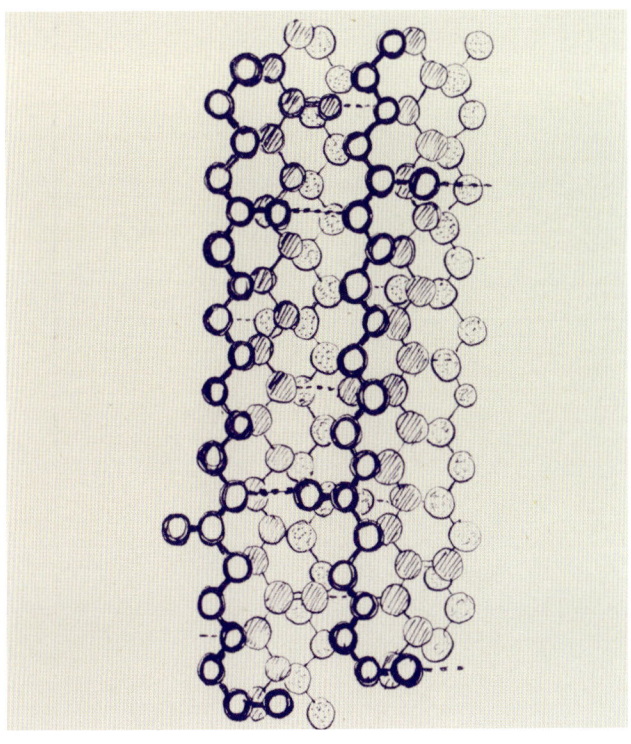

8.54 [c] Nylon diagram

ORTHOCLASE

$KAlSi_3O_8$

A tectosilicate crystalline mineral of cloudy appearance, also known as felspar or alkali feldspar, commonly found in granite. The most common mineral in the earth's crust, felspar is used as a raw material in ceramics and glass; also an ingredient in scouring powder.

Crystallographer: Dr. W.H. Taylor

Cavendish Laboratory, University of Cambridge

8.29 [1-2] Orthoclase

Ball-and-spoke structures, 8.29 [2] with tetrahedra

Source for 8.29 [1]: Bragg (1937), p.236, fig.131

Souvenir Book, p.9; *British Textiles*, April 1951, p.53

Collections: AAD 1977/3/553, WTA (dyelines 8.29 [1]); AAD 1977/3/553, V&A Circ. 78K-1968, WTA (dyelines 8.29 [2])

Diagram 8.29 [2] used by Old Bleach Linen Company

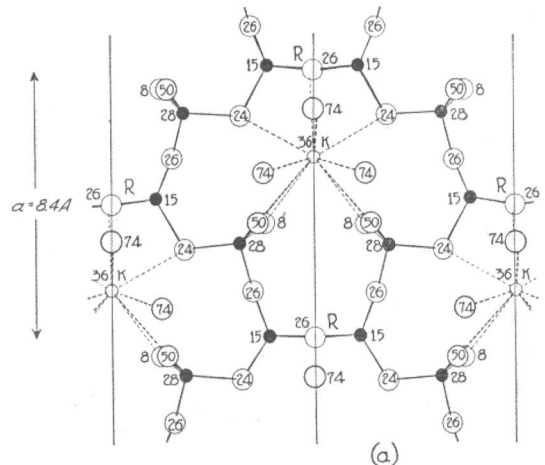

Source for Orthoclase 8.29 [1]

W.L. Bragg, Atomic Structure of Minerals, 1937, p.236, fig.131

8.29 [2] Orthoclase diagram

OXALIC ACID DIHYDRATE

$C_2H_2O_4$ $2H_2O$

Oxalic acid is a strong organic acid, usually synthesised in dihydrate form. Oxalic acid occurs naturally in many plants, including sorrel (Oxalis), rhubarb, spinach, cocoa and most nuts. Used as a mordant in dyeing; also used in wood restorers and rust-removing products.

Crystallographer: Professor W.H. Zachariasen

Department of Physics, University of Chicago

8.36 Oxalic acid dihydrate

Ball-and-spoke structure; 'With added outline which is more or less a packing diagram.'

Possible source: Bragg (1937), p.418, fig.15: 'This has been investigated by at least 5 sets of workers, who are all agreed on the general arrangement but differ as to the exact position of the atoms, and express their coordinates, in some cases, on different axes.'

Collections: WTA

Diagram possibly used later by A.C. Gill (AAD 1977/3/327)

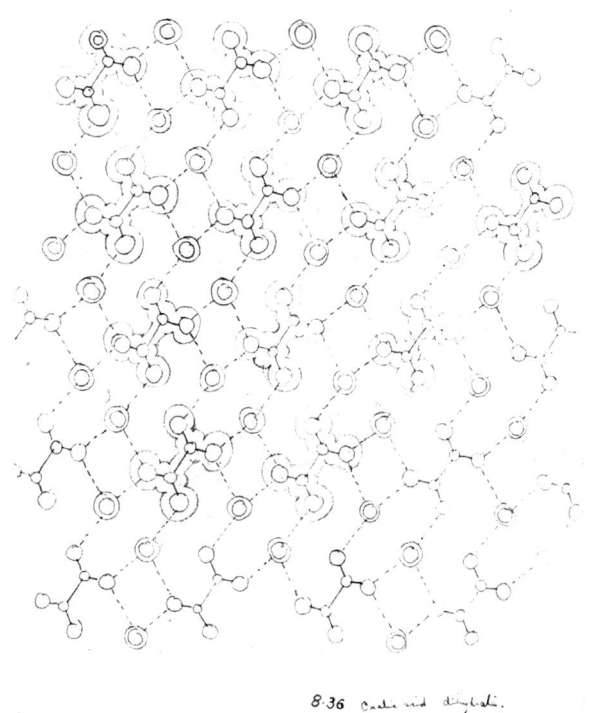

8.36 Oxalic acid dihydrate diagram

PARA DINITROPHENOL

$C_6H_4N_2O_5$ [dinitrophenol]

An industrially manufactured chemical compound known as pNP, closely related to dinitrophenol (DNP). DNP and pNP both act as metabolic uncouplers, speeding up the consumption of energy in cells. DNP was widely promoted as a dieting aid during the 1930s.

Crystal structure: 'Common knowledge'

8.21 Para dinitrophenol

Electron-density map

PENTAERYTHRITOL

$C(CH_2OH)_4$
An organic compound used as an agent in chemical reactions. Its main application is in the synthesis of Pentaerythritol tetranitrate (PETN), a high explosive patented by the German government in 1912, used in Semtex plastic explosive.
Crystallographer: Professor Gordon Cox
Department of Chemistry, University of Leeds

8.18 Pentaerythritol
Ball-and-spoke structure
Possible source: Llewellyn et al (1937), pp.883-894
Souvenir Book, p.10; British Textiles, p.52
Collections: AAD 1977/3/560 (notes); AAD 1977/3/561, V&A Circ.78M-1968 (dyelines); DCA 5384-3 (coloured dyeline);
Diagram used by James Templeton; Wood Brothers

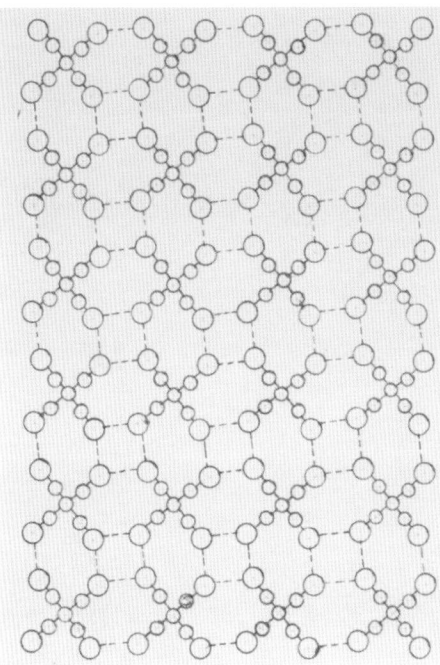

8.18 Pentaerythritol diagram

PEROVSKITE

$CaTiO_3$
Calcium titanium oxide: a naturally occurring hard, dark, lustrous crystalline mineral, named after Russian mineralogist, Count Lev Aleksevich von Perovski; also the name for a group of crystals with the same orthorhombic structure. Perovskites have useful electrical properties and can be made synthetically. They are important in the field of ferroelectricity where they are used as catalyst electrodes, sensors and superconductors.
Crystallograper: Dr. Helen Megaw
Cavendish Laboratory, University of Cambridge

8.1 Perovskite
Ball-and-spoke structure: 'Three different kinds of atoms in projection.'
Source: Megaw (1957), p.91, fig.5.3d; p.125, fig.6.1
Souvenir Book, p.0: 'The diagrams reveal that perovksite is close-packed and continuous in three dimensions'; *British Textiles*, p.53
Collections: AAD 1977/3/524 (coloured drawing); AAD 1977/3/524, WTA (dyelines)
Diagram used by James Templeton; Vanners & Fennell

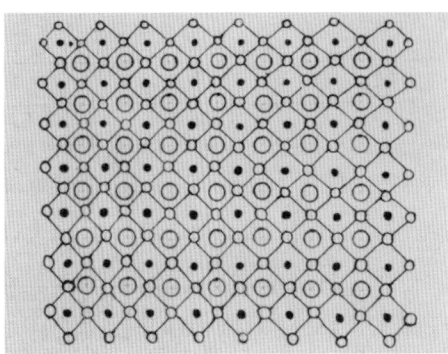

8.1 Perovskite diagram

POLYTHENE

$(C_2 – H_4)_n$ [general formula]; $C_{10}H_8O_4$ [polyethylene terephthalate]
A thermoplastic polymer, also known as polyethylene, consisting of long chains of ethylene molecules. Discovered by accident in 1898 by a German chemist, Hans Pechmann, when heating diazomethane. Subsequently developed by chemists at ICI's Alkali Division at Northwich from 1933; manufactured from 1939 by synthesising ethylene and benzaldehyde under high pressure. One of the most stable and inert polymers, polythene is used for a diverse range of packaging, both rigid and flexible. Categorised according to density (high, medium and low), its applications include detergent bottles, cling film, plastic bags and hoses.
Crystallographer: Charles William Bunn
ICI (Alkali Division), Northwich; ICI (Plastics Division), Welwyn Garden City

8.59 [a-e] Polythene
Electron-density maps: 8.59a-b: 'Different representations of molecule, which is indefinitely long; (a) and (b) show it from different directions.' 8.59c-d: 'Repeats indefinitely in length.' 8.59e: 'Must be used as a continued repeat, and not an isolated motif. Some contours may be omitted, if done throughout, but not too many.'
Source (8.59a-d): Bunn (1939), p.489, fig.5
Source (8.59e): Bunn (1939), p.489, fig.4
Souvenir Book, p.0 (8.59c, e), p.7 (8.59d); *British Textiles*, p.52
Collections: AAD 1977/3/467, V&A T.446F-1977 (dyelines 8.59c-d); AAD 1977/3/468 (dyeline 8.59e); AAD 1977/3/469, V&A Circ.78C-1968 (dyelines 8.59a-b)
Diagrams used by Elkington (8.59d); Dobroyd (8.59c); A.C.Gill (8.59c); Goodearl Brothers (8.59b); London Typographical Designers (8.59c)

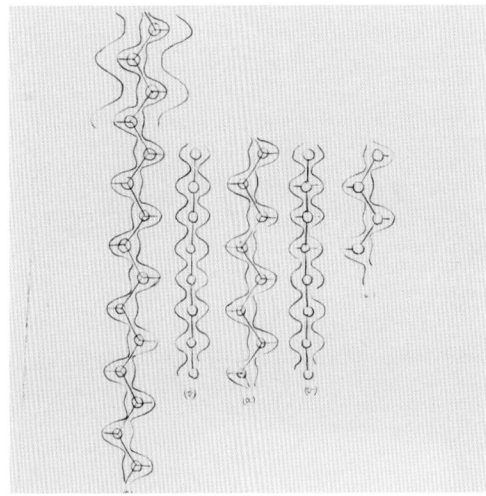

8.59 [a-b] Polythene diagram

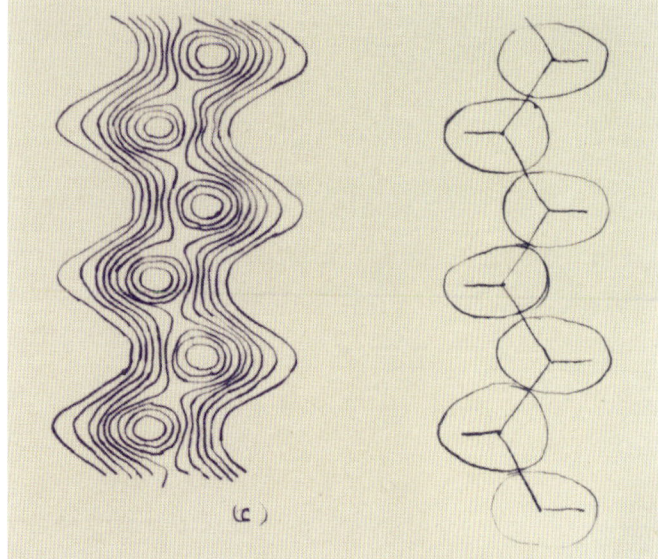

8.59 [c-d] Polythene diagram

8.59 [e] Polythene diagram

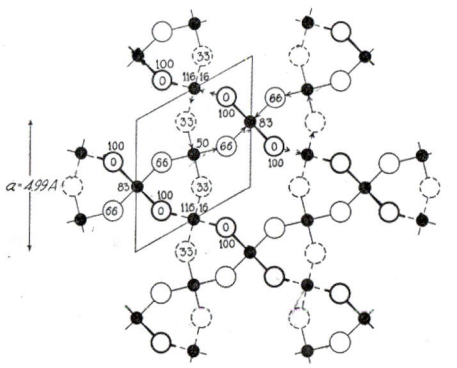

Source for Quartz 8.10
W.L. Bragg, Atomic Structure of Minerals, 1937, p.84, fig.47

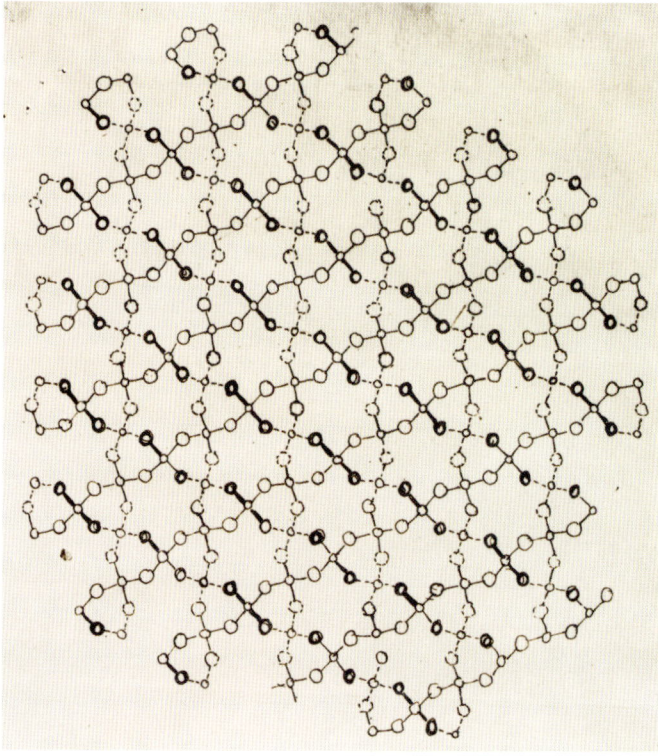

8.10 Quartz diagram

QUARTZ

SiO_2

One of the most common minerals, quartz (silicon dioxide) is a tectosilicate, vitreous and crystalline in appearance. Some types of quartz have pronounced rhombohedral crystals; others are microcrystalline. In its pure form, known as rock crystal, quartz is colourless; but it also occurs in a variety of colours, including purple (amethyst), pink (rose quartz), white (chalcedony), orange-red (carnelian), and multi-coloured (agate and onyx). Other colours can be created through heat treatment.

Crystallographers: Sir William Henry Bragg / Professor Lawrence Bragg

Royal Institution, London / Cavendish Laboratory, University of Cambridge

8.10 Quartz

Ball-and-spoke structure: 'Two kinds of atoms (large and small circles). Different thicknesses of lines indicate different heights.'

Sources: Bragg, Gibbs (1925), p.421, fig.13; Bragg (1937), p.84, fig.47

Souvenir Book, p.11

Collections: AAD 1977/3/491-493 (annotated drawings); AAD 1977/3/525, WTA (dyeline and photostat)

Diagram used by James Templeton

RESORCINOL

$C_6H_4(OH)_2$ [general formula]

An organic compound, also known as resorcin, which crystallises from benzene. A hydrocarbon, resorcinol can be synthesised by combining various dihydroxy phenols with acids. It can also be distilled from Brazilwood extract, or created by fusing certain natural resins with potassium hydroxide. Widely used as an antiseptic and disinfectant; also an ingredient in various skin treatments for acne, eczema and psoriasis

Crystallographer: Professor John Monteath Robertson

Department of Chemistry, University of Glasgow

8.17 Resorcinol

Ball-and-spoke structure: 'Atoms at six corners of hexagon are of different kind from those shown as circles. Full and dotted lines are strong and weak bonds.'

Related source: Monteath Robertson (July 1948), p.105, fig. 6

Souvenir Book, p.10

Collections: AAD 1977/3/562 (photostat)

Diagram used by James Templeton

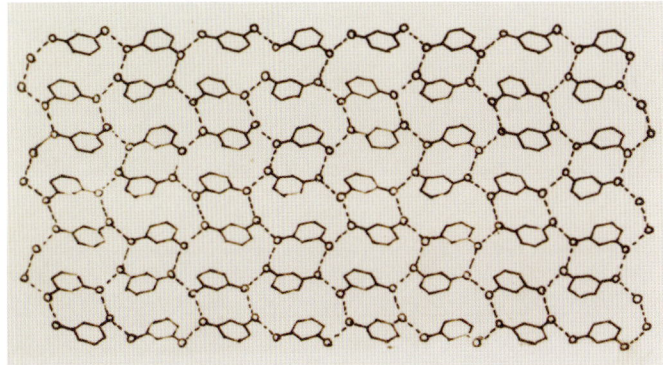

8.17 Resorcinol diagram

RUBIDIUM DITHIONATE

$Rb_2S_2O_6$

A compound incorporating rubidium, a soft silvery-white alkali metal found in the mineral lepidolite. Highly flammable in air, rubidium is sometimes added to fireworks to produce a purple colour.

Crystallographer: Andre V. Wendling

McGill University, Montreal

8.13 Rubidium dithionate

Source: Wendling (1938)

SPINEL

$MgAl_2O_4$

A family of minerals, often composed of octahedral crystals, including several semi-precious stones. True spinel is sometimes colourless, but can be blue, black, brown, green, purple, red or white. Spinels fall into two main groups: aluminium-based and iron-based. One type, ringwoodite, is buried deep in the earth's mantle.

Crystallographer: W.H. Bragg

Royal Institution, London

8.14 Spinel

Ball-and-spoke structure: 'Three kinds of atoms. Dotted and full lines at different height.'

Source: Bragg (1915), p.305; Bragg (1937), p.99, fig.58

Collections: WTA (dyeline)

Diagram used by Wedgwood

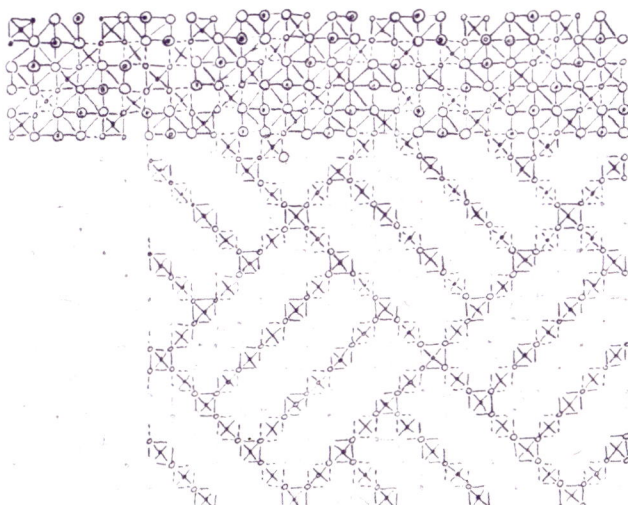

8.14 Spinel diagram

ZINC HYDROXIDE

$Zn(OH)_2$

Hydrated zinc: an inorganic chemical compound derived from the mineral zincite. Takes the form of white crystalline powder; used as an absorbing agent in chemical processes and surgical dressings.

Crystallographer: Dr. Helen Megaw

Cavendish Laboratory, University of Cambridge

8.39 Zinc hydroxide

Ball-and-spoke structure

Source: Bernal, Megaw (1935), p.398, fig.5

Souvenir Book, p.11

Collections: AAD 1977/3/548, WTA (dyelines initialled 'H.D.M.')

Diagram used by Carter & Co.

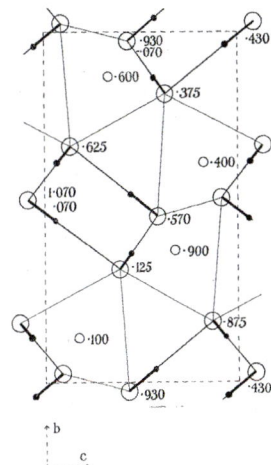

Source for Zinc hydroxide 8.39

J.D. Bernal, H.D. Megaw, 'The Function of Hydrogen in Intermolecular Forces', Proceedings of the Royal Society, Series A, 2 September 1935, vol.151, no.873, p.398, fig. 5;

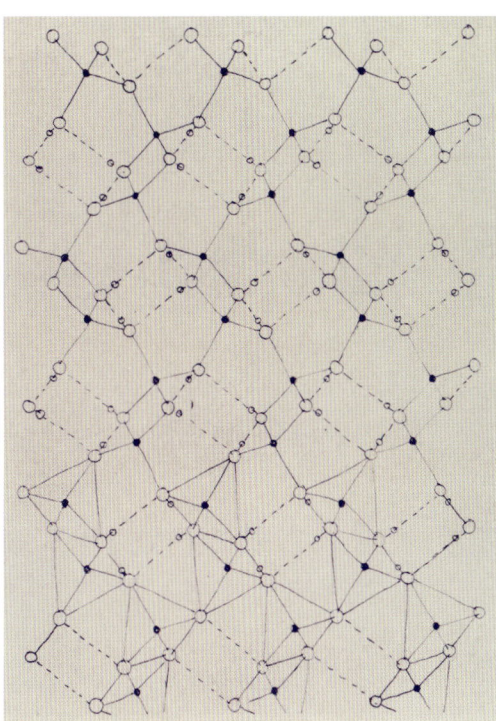

8.39 Zinc hydroxide diagram

CHAPTER SIX

CRYSTALLOGRAPHERS: THE SCIENTISTS BEHIND THE DESIGNS

PROFESSOR JOHN DESMOND BERNAL (1901-1971)

Highly respected, but politically controversial, physicist and crystallographer, nicknamed 'Sage', whose pioneering research into amino acids, sterols, peptides and proteins laid the foundations for major post-war breakthroughs in molecular biology. Studied mathematics, physics, chemistry, geology and mineralogy at Emmanuel College, Cambridge (1919-23), Initially worked at Royal Institution under W.H. Bragg, where he researched the crystal structure of metal alloys and developed an important piece of crystallographic equipment: the X-ray photogoniometer. Lecturer in structural crystallography at Cambridge from 1927; research students included Helen Megaw and Dorothy Hodgkin. Assistant Director of Research at Cavendish Laboratory (1935-38). Chair of Physics at Birkbeck College, University of London from 1938. Author of *The Social Function of Science* (1937), *Science in History* (1954) and *Marx and Science* (1957).

Festival Pattern Group: Not directly involved but his joint research with Helen Megaw was used as a source. FPG 49 was based on Bernal chart used for interpreting oscillation photographs of crystals. Curated 'Great Scientific Discoveries and their Applications' at the Exhibition of Science (*Science Guide*, pp.42-45).

SIR WILLIAM HENRY BRAGG (1862-1942)

World-renowned physicist and pioneer of X-ray crystallography. Studied mathematics at Trinity College, Cambridge (1881-85). Lecturer in applied mathematics at University of Adelaide from 1886. Began experimenting with X-rays in 1895; conducted ground-breaking research on alpha particles and radiation. Appointed Cavendish Professor of Physics at University of Leeds in 1909. Published *Studies in Radioactivity* (1912). Collaborated with his son, Lawrence Bragg, developing applications for X-ray photography in crystallography (1912-15); also invented X-ray spectrometer. Co-wrote *X-Rays and Crystal Structure* (1915). Jointly awarded Nobel Prize for Physics (1915). Appointed Quain Professor of Physics at University College, London shortly afterwards. Knighted in 1920. Director of Royal Institution from 1923; research team included J.D. Bernal, Gordon Cox, Kathleen Lonsdale and J.M. Robertson. Published *The World of Sound* (1920) and *The Nature of Things* (1925).

Festival Pattern Group Crystal Structures: **Quartz 8.10; Spinel 8.14** Authorised posthumously by Lawrence Bragg.

SIR WILLIAM LAWRENCE BRAGG (1890-1971)

Internationally celebrated pioneer of X-ray crystallography. Studied mathematics under his father, W.H. Bragg, at University of Adelaide (1905-08), and physics at Trinity College, Cambridge (1909-12). After seeing Max von Laue's X-ray diffraction photographs of crystals in 1912 he formulated the theory that X-ray waves are deflected by the planes of atoms in crystalline materials, so the shadow patterns in X-ray diffraction photographs indirectly reveal the atomic structure of matter (Bragg's Law). Based in Cambridge, but worked closely with his father in Leeds from 1912-15 developing the theory and practice of X-ray crystallography outlined in *X-Rays and Crystal Structure* (1915); jointly awarded Nobel Prize for Physics in 1915. Used X-ray spectrometer to explore crystal structures of sodium and potassium chloride, mica and diamond. Appointed Langworthy Professor of Physics at Manchester University (1919-38); tackled increasingly ambitious structures, particularly silicates, adopting Fourier methods to calculate electron-density distribution. Research group included G.W. Brindley, W.W. Jackson, W.H. Taylor and J. West. Published *The Crystalline State* (1933) and *Atomic Structure of Minerals* (1937). Appointed Professor of Physics at Cambridge in charge of Cavendish Laboratory (1938-54). Crystallography team included Helen Megaw, W.H. Taylor, Max Perutz and John Kendrew. Knighted in 1941. Fullerian Professor at the Royal Institution (1954-66).

Festival Pattern Group Crystal Structures: **Beryl 8.9; Quartz 8.10**

'Miss Megaw showed me yesterday the samples of the application of crystal patterns to industrial design. I must write to say how delighted I am that the patterns have been applied for this purpose... The patterns she showed me yesterday are the practical realization of what we have long wished to see. Since my father and I started this work in 1912 the pupils whom he and I have trained have produced a huge series of these complex patterns, so there is plenty more material for you should you need it. I hope they earn us some dollars.' (Letter to MHT, 11/5/1951. DCA 5384-4).

PROFESSOR GEORGE W. BRINDLEY (1905-1983)

Physicist and crystallographer. Studied physics at Manchester University under Lawrence Bragg; completed MSc (1928), followed by PhD in Leeds (1933). Continued his research at Leeds, focusing on the scattering of X-rays by atoms; appointed Reader in X-ray Physics (1948). Began to study clay minerals during mid 1940s; published *Crystal Structure of Clay Minerals and their X-ray Identification* (1951). Appointed Research Professor at Pennsylvania State College (later University) in 1953, subsequently Chair of Ceramic Technology (1955-62), latterly Professor of Mineral Sciences (1962-73). Awarded Roebling Medal by Mineralogical Society of America (1970). Nickel-bearing clay Brindleyite named in his honour.

Festival Pattern Group Crystal Structures: **China clay 8.2, 8.3, 8.6, 8.7; Clay minerals 8.55**

'It is with great pleasure that I fall in with your wishes. I only wish I had many more suitable designs for you at £5 a time! / I enclose one, however, of displaced SiO nets seen in projection & elevation. It is a diagram which will be sent to Acta [Crystallographica] in a few days time with a paper on chlorites.' (Letter to HM, 1/3/50. AAD1977/3/725). 'I think that in any printed statement regarding the origins of these designs, it should be made clear that the manufacturer has used the scientific diagrams "as the basis of a design"' (Letter to HM, 19/2/1951. AAD1977/3/726).

DR. JUNE M. BROOMHEAD

Talented crystallographer, highly regarded by Dorothy Hodgkin, who published important papers on purines and pyrimidines (1948-51). Studied at Cambridge, completed PhD at Cavendish Laboratory (1949). Analysed structures of adenine and guanine hydrochloride; her results were later used by Crick and Watson when working on

DNA. Research assistant to Dorothy Hodgkin in Oxford (1949-50); processed important data on Vitamin B12 using X-RAC computer at Penn State College (1950). Gave up career after marrying and moving to US; later settled in Canada (married name Lindsey). *Festival Pattern Group Crystal Structures:* **Adenine hydrochloride 8.47, 8.48; Guanine hydrochloride 8.49, 8.50**

DR. CHARLES WILLIAM BUNN (1905-1990)

Chemist and crystallographer specialising in long-chain polymers. Studied chemistry at Exeter College, Oxford (1923-27). Spent most of career working for ICI, initially at Alkali Division at Winnington, Northwich, Cheshire (1928-46). Largely self-taught in crystallography; developed links with Lawrence Bragg in Manchester during 1930s. Collaborated with Dorothy Hodgkin at Oxford on wartime research into structure of penicillin (1944-45). Published *Chemical Crystallography* (1946). Worked at ICI's Plastics Division at Welwyn Garden City (1946-63). Dewar Research Fellow at Royal Institution (1963-72). Published *Crystals: their role in nature and in science* (1964).

Festival Pattern Group Crystal Structures: **Polythene 8.59; Nylon 8.54**

'Mr Bunn has not signed any agreements because the negotiations went straight through I.C.I.' (HM's notes. AAD1977/3/93).

SIR ERNEST GORDON COX (1906-1996)

Crystallographer and civil servant. Studied physics at Bristol University (1924-27). Joined W.H. Bragg's laboratory at the Royal Institution in 1927 as research assistant. Determined the structure of benzene using a low temperature camera he designed himself. Lecturer at Birmingham University from 1931; Reader in Crystallography from 1940; worked on crystal structures of sugars, vitamin C and pentaerythritol (relevant to his wartime work on explosives). Professor of Inorganic and Structural Chemistry at University of Leeds (1945-60); actively promoted the use of computers in crystallography and built up a dynamic research team known as Cox's Pippins. Secretary to the Agricultural Research Council (1960-71). Knighted in 1964.

Festival Pattern Group Crystal Structure: **Pentaerythritol 8.18**

PROFESSOR DOROTHY M. CROWFOOT HODGKIN (1910-1994)

Nobel Prize-winning crystallographer who determined the structures of penicillin, vitamin B_{12} and insulin. Studied chemistry at Somerville College, Oxford (1928-32). Research student in J.D. Bernal's X-ray crystallography laboratory in Mineralogy Department, Cambridge (1932-34). Took X-ray diffraction photographs of the enzyme pepsin; established that protein crystals should be studied in their mother liquid rather than air-dried form. Fellow (and later tutor) at Somerville College, Oxford from 1934 onwards; completed PhD on sterols (1936). Established crystallography laboratory in University Museum at Oxford. Embarked on ground-breaking research into the structure of insulin; took first X-ray photographs of insulin crystals in 1934. Conceived important theory that the way to solve complex structures was to study an isomorphous crystal, one in which a normal atom was replaced by a heavier atom. Married Thomas Hodgkin in 1937 but published under her maiden name Crowfoot until 1949. Worked on penicillin from 1942, collaborating with C.W. Bunn from ICI from 1944; determined its structure in 1945, but their discovery was kept

secret by the government until 1949. Lecturer and demonstrator in Department of Chemical Crystallography at Oxford from 1946; later appointed Reader. Began researching vitamin B_{12} in 1948, pioneering the use of Patterson maps and computers; determined its structure in 1955. Appointed Wolfson Research Professor (1960), a new post endowed by the Royal Society. Awarded Nobel Prize for Chemistry (1964) and Order of Merit (1965). Continued working on insulin, finally solved its full structure in 1969. President of the International Union of Crystallography (1972-77).

Festival Pattern Group Crystal Structures: **Insulin 8.24, 8.25, 8.27**

'I've lost your [FPG authorisation] form – accidentally on purpose, I suspect. I didn't at all like the idea of signing it. I don't propose to talk about your project but I should hate to promise secrecy... I don't mind giving your firms permission to use the insulin patterns... I feel rather doubtful whether I own any copyright of a pattern perpetrated by nature. But I suspect that if I do, I shouldn't sell it for £5. But I think it's a nice idea! I hope your friends go ahead, even without proper copyright declarations.' Letter to HM, 1/3/1950. AAD1977/3/723).

SIR JOHN C. KENDREW (1917-1997)

Pioneer molecular biologist and crystallographer. Studied science at Trinity College, Cambridge (1936-39), specialising in chemistry. Met J.D. Bernal during the war who urged him to apply X-ray crystallography to molecular biology. Joined Cavendish Laboratory in 1945 on an ICI research fellowship; began working on protein structures alongside Max Perutz. Part of new Medical Research Council Unit for the Study of the Molecular Structure of Biological Systems from 1947. Completed PhD comparing foetal and adult haemoglobin (1949). Started researching myoglobin, using EDSAC I –II computers to process data. Created first molecular model of myoglobin (known as 'the sausage') in 1957 - the first protein structure determined using crystallography – followed by a higher resolution model known as the 'forest of rods' (1959). Awarded Nobel Prize for Chemistry in 1962. Deputy Chairman of Laboratory of Molecular Biology and Head of Structural Studies Division from 1962. Fellow of Peterhouse (1947-97); part-time Reader at Royal Institution (1954-68); scientific adviser to government (1961-74). Knighted in 1974. Director-General of European Molecular Biology Laboratory, Heidelberg (1975-82).

Festival Pattern Group Crystal Structures: **Whale myoglobin 8.46 [a-b]; Horse myoglobin 8.46 [c-g]**

'I understand that your firm is manufacturing products with crystal structure patterns, in connection with the Festival of Britain. Some of these patterns – those called "horse myoglobin" and "whale myoglobin" – were submitted by myself and I should be most interested to have the opportunity of purchasing small amounts of materials using these designs.' (Letter to John Line, redirected to ICI Leathercloth, 13/4/1951. DCA 5384-4).

DAME KATHLEEN LONSDALE (1903-1971)

Pioneer crystallographer. Studied physics at Bedford College, London (1919-22). Took up research post with W.H. Bragg in 1922, initially at University College, London, subsequently at Royal Institution. Determined the structure of hexamethylbenzene while working in Leeds (1927-29). Returned to RI in 1931; worked on magnetic anisotrophy of crystals and thermal movement of atoms in crystals. Appointed Reader in Crystallography at University College

(1946); Professor of Chemistry in Department of Crystallography (1949-68). Later research fields included synthetic diamonds and endemic bladder stones.

Festival Pattern Group: Although none of Lonsdale's crystal structures were used, she was instrumental in triggering the FPG: in the first place it was her lecture on crystallography to the Society of Industrial Artists in May 1949 that prompted Mark Hartland Thomas from the COID to contact Helen Megaw.

DR. HELEN DICK MEGAW (1907-2002)

Highly respected crystallographer who initiated the Festival Pattern Group. Studied science at Queen's University, Belfast (1925-26), and Girton College, Cambridge (1926-30). Joined J.D. Bernal's laboratory in Cambridge as a research student (1930); worked alongside Dorothy Crowfoot (1932-34); completed her PhD (1934). Early research focused on the crystal structure of ice; Megaw Island in the Antarctic was later named after her. Co-wrote an article with J.D. Bernal on 'The Function of Hydrogen in Intermolecular Forces' (1935), later used as a source for FPG designs. Spent a year in Vienna working with the chemist Professor Hermann Mark (1934-35), followed by a year at the Clarendon Laboratory, Oxford (1935-36). Schoolteacher in Bedford and Bradford (1936-43). Joined Philips Lamps (1943); determined crystal structure of barium titanate, a type of perovskite, which triggered her interest in ferroelectrics. Worked with Bernal at Birkbeck College, London (1945-46). Joined Cavendish Laboratory, Cambridge (1946), where her colleagues included Sir Lawrence Bragg, Max Perutz, John Kendrew and W.H. Taylor. Appointed Assistant Director of Research (1949); also lecturer in chemistry from 1959. Fellow of Girton College, and Director of Studies in Physical Science (1946-72). Remained at the Cavendish until her retirement in 1972, focusing on minerals, especially perovskites and feldspars, and inorganic materials. Published *Ferroelectricity in Crystals* (1957) and *Crystal Structures: A Working Approach* (1973).

Festival Pattern Group Crystal Structures: **Afwillite 8.37, 8.38, 8.42, 8.44, 8.45; Aluminium hydroxide Hydrargillite 8.33; 8.8, 8.33; Boric acid 8.34; Brucite 8.32; Perovskite 8.1, Zinc hydroxide 8.39**

Initiator of the Festival Pattern Group: Adviser on Crystal Structure Diagrams for the Festival of Britain Exhibition (October 1949 to May 1951).

DR. MAX FERDINAND PERUTZ (1914-2002)

Nobel prize-winning molecular biologist and crystallographer. Born in Vienna, the son of a textile manufacturer. Studied chemistry at University of Vienna (1932-36). PhD at Cavendish Laboratory, Cambridge (1936-40). Worked with J.D. Bernal initially, who encouraged him to use X-ray crystallography to work out the structure of proteins, regarded by Perutz as 'the secret of life.' Embarked on a lifetime of research into haemoglobin in 1937. Interned for nine months during World War II. ICI Research Fellow (1945-46). Director of Medical Research Council Unit for the Study of the Molecular Structure of Biological Systems from 1947, renamed Molecular Biology Research Unit in 1957. (This was where James Watson and Francis Crick discovered the double helix structure of DNA in 1953). Developed vital technique of isomorphous replacement, whereby heavy atoms such as mercury were 'tied' to normal atoms, thus enabling crystals with and without heavy atoms to be compared. This resolved the problem of 'phasing', and paved

the way for protein structures to be solved. The first to benefit was his colleague John Kendrew, who determined the structure of myoglobin (1957-59). Perutz worked out the complex structure of haemoglobin in 1959. Awarded Nobel Prize for Chemistry (1962). Published *Proteins and Nucleic Acids: Structure and Function* (1962). Chairman of newly established Laboratory of Molecular Biology from 1962, which brought together his research group with those of biochemist Fred Sanger and virus researcher Aaron Klug. Continued exploring other aspects of haemoglobin for several decades, working out the structures of oxy- and deoxy-haemoglobin, and researching mutations leading to sickle cell anaemia. Retired as chairman of LMB in 1979, but continued his research there until 1995, latterly focusing on neurodegenerative diseases. Published two collections of essays: *Is Science Necessary?* (1989) and *I Wish I'd Made You Angry Earlier* (1998).

Festival Pattern Group Crystal Structures: **Haemoglobin 8.26; Horse Methaemoglobin 8.23**

PROFESSOR JOHN MONTEATH ROBERTSON (1900-1989)

Leading chemical crystallographer. Studied science at University of Glasgow (1919-23), specialising in chemistry; completed PhD (1926). Went to Royal Institution to study with W.H. Bragg, working alongside J.D. Bernal and Kathleen Lonsdale on the structure of organic molecules. Appointed senior lecturer at University of Sheffield (1939). Gardiner Professor of Chemistry at Glasgow from 1942; Director of the Chemical Laboratories (1955-70).

Festival Pattern Group Crystal Structures: **Hexamethylene diamine 8.40, 8.41, 8.51; Resorcinol 8.17**

'The idea which you put forward is a very interesting one, and I would certainly like to see some of these patterns fully worked out. I believe the effects might be quite striking. I find it slightly amusing to think of these elaborate calculations achieving such a utilitarian outlet.' (Letter to HM, 10/2/1950. AAD1977/3/720).

DR. W.H. TAYLOR (1905-1984)

Leading physicist and crystallographer. Studied physics at Manchester University under Lawrence Bragg, graduating in 1926. PhD student and lecturer at Manchester until 1934. Determined the crystal structures of numerous silicate minerals during late 1920s, mainly published in *Zeitschrift für Kristallographie*, followed by apophyllite (1931), zeolite (1933) and orthoclase (1933-34). Leverhulme Research Fellowship (1934-36), initially working with J.D. Bernal at Cambridge, and then at Royal Institution under W.H. Bragg, branching out into organic structures. Head of Physics at Manchester College of Technology (1936-45). Reader in Crystallography at Cavendish Laboratory, Cambridge (1945-71), working alongside Helen Megaw; continued research on feldspars, as well as metals and alloys. Instrumental in founding International Union of Crystallography (1946). Edited *Philosophical Magazine* (1971-76). Awarded Roebling Medal by Mineralogical Society of America (1979).

Festival Pattern Group Crystal Structures: **Apophyllite 8.30, 8.31; Orthoclase 8.29**

Chapter 1 - Notes

1. Letter from HM to J.R.M. Brumwell, 20/2/1946 (AAD 1977/3/12)

2. Letter from J.R.M. Brumwell to HM, 1/3/1946 (AAD 1977/3/13).

3. Transcript of letter from Barbara Hepworth to J.R.M. Brumwell, forwarded to HM, 1/3/1946 (AAD 1977/3/15).

4. Helen Megaw, 'Pattern in Crystallography', unpublished essay, November 1946, henceforth referred to as Megaw (1946). See Chapter 4 for edited transcript.

5. Ibid.

6. Ibid.

7. Letter from Kathleen Lonsdale to HM, 16/11/1948 (AAD 1977/3/28).

8. Letter from MHT to HM, 1/6/1949 (AAD 1977/3/31).

9. Letter from HM to MHT, 7/6/1949 (AAD 1977/3/32).

10. Letter from MHT to HM, 20/6/1949 (AAD 1977/3/33).

11. Mark Hartland Thomas, *The Souvenir Book of Crystal Designs*, COID/HMSO, 1951, pp.3-5.

12. Letter from HM to Miss O'Donovan, COID, 8/12/1949 (AAD 1977/3/60).

13. For published scientific sources of FPG diagrams, see Chapter 5: An A-Z of Crystal Structures.

14. In the Crystal Structure Diagram authorisation form completed and signed by Max Perutz, March 1950, he describes it as: 'Patterson of reduced haemoglobin. Unpublished.' (DCA 9466). According to Joyce Baldwin, a former colleague, the diagram probably depicts horse haemoglobin: 'This form of haemoglobin was the only one that had produced crystals in a space group with three-fold symmetry. I expect that studies on this particular type of haemoglobin crystal were not continued very far as the large length of the c-axis would have made it very difficult to work with at the time. However, the hexagonal symmetry of the Patterson map would have made it an attractive choice for the 1951 exhibition. Later work on horse reduced haemoglobin was done on crystals of a different space group, so this particular map may never have been included in a publication.' (Email to author, 17/1/2008).

15. Letter from HM to Sophie Forgan, 26/5/1997. I am indebted to Sophie Forgan for providing access to her correspondence with HM.

16. Letter from HM to Dorothy Hodgkin, 30/1/1950 (AAD 1977/3/721).

17. 'My recollections of my connection with the Crystallography theme in the 1951 Festival of Britain as written down in 1993 by Helen D. Megaw', transcript of notes sent by HM to Sophie Forgan, 26/5/1997, henceforth referred to as Megaw (1993).

18. Megaw (1946), op. cit.

19. Letter from J. Weyman, COID, to HM, 6/1/1950 (DCA 9466).

20. Letter from HM to Professor John Monteath Robertson, 20/1/1950 (AAD 1977/3/718).

21. Letter from Dorothy Hodgkin to HM, 1/3/1950 (AAD 1977/3/723).

22. Letter from HM to MHT, 21/3/1951 (DCA 5395).

23. Letter from G.W. Brindley to HM, 19/2/1951 (AAD 1977/3/726).

24. Megaw (1993), op. cit.

25. The artist Richard Hamilton was deeply preoccupied with biological imagery at this date. In July 1951 he mounted an installation at the ICA called Growth and Form exploring the relationship between art and science. However, although Hamilton drew liberally on scientific material, he did not actively collaborate with scientists, so there appear to be no direct parallels with the FPG.

26. Minutes of first FPG Meeting, 16/1/1949 (DCA 5396).

27. Ibid.

28. Megaw (1946), op. cit.

29. Helen Megaw, 'Notes on Crystal Structure Diagrams', January 1950. See Chapter 4.

30. See Chapter 3: Catalogue of Festival Pattern Group Manufacturers and Designs.

31. File note by MHT, 4/11/1949 (DCA 5384-1).

32. Minutes of third FPG Meeting, 16/3/1950 (DCA 5396).

33. Megaw (1993), op. cit.

34. Letter from HM to MHT, 12/8/1950 (DCA 5384-2).

35. Minutes of fourth FPG Meeting, 13//7/1950 (DCA 5396).

36. FPG Report of Progress to COID, 30/10/1950 (DCA 5396).

37. Letter from Patience Clifford to MHT, 8/1/1951 (DCA 5384-3).

38. Note from Gordon Russell to MHT, 17/4/1951 (DCA 5384-4).

39. Letter from Misha Black to MHT, 1/9/1950 (DCA 5384-2).

40. *Science Guide*, p.17.

41. *South Bank Guide,* p.60-62.

42. An internal memo from R. Marsden to MHT, 7/12/1950 lists the manufacturers. No photographs have so far come to light, but the display is mentioned in the FPG graphics panel illustrated in the *Souvenir Book* (FPG 44). Another display of FPG products planned for the Exhibition of Science was dropped at a late stage due to unspecified 'difficulties'. Letter from MHT to Ian Cox, 28/2/1951. (DCA 5384-3).

43. 'Crystal Patterns for Fabrics', *The Times*, 24 April 1951.

44. 'The "Atomic" Tie', *East Anglian Daily Times*, 16 April 1951.

45. Letter from MHT to FPG manufacturers, 17/5/1951 (DCA 5396).

46. Bernard Hollowood, 'Petri's Kaleidoscope', *Punch*, 8 August 1951, pp.167-168.

47. Letter from HM to W. Haigh, Dobroyd, 12/5/1951 (AAD 1977/3/208).

48. Letter from Sir Lawrence Bragg to MHT, 11/5/1951 (DCA 5384-4).

49. Letter from MHT to Sir Lawrence Bragg, 18/5/1951 (DCA 5384-4).

50. Nottingham Castle Museum NCM 1962-43.

51. Letter from HM to MHT, 4/7/1951 (DCA 5384-4).

52. V&A Circ. T.368-1977.

53. Letter from MHT to J.R.M. Brumwell, 19/7/1951 (DCA 5384-4).

54. Letter from Hogg & Mitchell to MHT, 16/7/1951 (DCA 5384-4).

55. Correspondence between Oxvar and HM, 1951-55 (AAD 1977/3/186-308).

56. Whitworth Art Gallery T.1992.35 and V&A T.117.182-1992.

57. Minutes of final FPG Meeting, 16/10/1951 (DCA 5396).

58. *Design*, October 1973, p.97.

59. Letter from HM to Sophie Forgan, 3/4/1997.

60. Megaw (1993), op. cit.

61. Letter from J.W. Chance to MHT, 5/10/1951 (DCA 5384-4).

62. Letter from D.C. Spicer to MHT, 17/4/1951 (DCA 5384-4).

63. The company donated production samples of 'Helmsley' in five colourways to the Whitworth Art Gallery in 1961.

64. Letter from J.A. Dunkin Wedd to HM, 14/4/1955 (AAD 1977/3/331). Warerite's FPG laminates were used on RMS Oronsay: *Design*, August 1951, p.16. See also, Paul Reilly, 'A decorative future for plastics laminates', *Design*, December 1954, pp.12-13.

65. Megaw (1993), op. cit.

66. Correspondence between J.A. Dunkin Wedd and HM, 1955 (AAD 1977/3/328-331).

67. J.A. Dunkin Wedd, *Pattern and Texture – Sources for Design*, Studio Publications, 1956, pp.79-85.

68. Letter from W.A. Dickie to MHT, 4/6/1951 (DCA 5384-4).

Bibliography A: Festival Pattern Group Crystal Structures – Scientific Sources

F.A. Bannister (1935), 'The crystal-structure of bismuth oxyhalides', *Mineralogical Magazine*, June 1935, vol.24, no.49, pp.49-58 [bismuth oxychloride]

M. Bailey (1949), 'The Crystal Structure of Diethyl Terephthalate', *Acta Crystallographica*, April 1949, vol.2, pp.120-126

J.D. Bernal, H.D. Megaw (1935), 'The Function of Hydrogen in Intermolecular Forces', *Proceedings of the Royal Society, Series A*, 2 September 1935, vol.151, no.873, pp.384-420

J. Boyes-Watson, E. Davidson, M.F. Perutz (1947), 'An X-ray study of horse methaemoglobin. I', *Proceedings of the Royal Society, Series A*, 26 September 1947, vol.191, no.1024, pp.83-132

W.P. Binnie, J.M. Robertson (April 1949), 'The Crystal Structure of Hexamethylenediamine and its Dihalides. Hexamethylenediamine Dihydrobromide', *Acta Crystallographica*, April 1949, vol.2, pp.116-120

W.P. Binnie, J.M. Robertson (June 1949), 'The Crystal Structure of Hexamethylenediamine Dihydrochloride', *Acta Crystallographica*, June 1949, vol.2, pp.180-188

W.H. Bragg (1915), *Philosophical Magazine*, 1915, no.30, p.305 [spinel]

W.H. Bragg, R.E. Gibbs (1925), 'The Structure of α and β Quartz', *Proceedings of the Royal Society, Series A*, 1925 vol.109, no.751, pp.405-427

W.L. Bragg, J. West (1926), 'The Structure of Beryl, $Be_3Al_2Si_6O_{18}$', *Proceedings of the Royal Society, Series A*, 1926, no.111, p.691

W.L. Bragg (1937), *Atomic Structure of Minerals*, Cornell University Press, New York / Oxford University Press, 1937

W.L. Bragg, J.C. Kendrew, M.F. Perutz (1950), 'Polypeptide Chain Configurations in Crystalline Proteins', *Proceedings of the Royal Society, Series A*, 10 October 1950, vol.203, no.1074, pp.321-357

W.L. Bragg, E.R. Howells, M.F. Perutz (January 1952), 'Arrangement of Polypeptide Chains in Horse Methaemoglobin', *Acta Crystallographica*, January 1952, vol.5, pp.136-141

W.L. Bragg, M.F. Perutz (March 1952), 'The External Form of the Haemoglobin Molecule. I,' *Acta Crystallographica*, March 1952, vol.5, pp.277-283

W.L. Bragg, M.F. Perutz (May 1952), 'The External Form of the Haemoglobin Molecule. II,' *Acta Crystallographica*, May 1952, vol.5, pp.323-328

W.L. Bragg, M.F. Perutz (July 1952), 'The Structure of Haemoglobin', *Proceedings of the Royal Society, Series A*, 22 July 1952, vol.213, no.1115, pp. 425-435

W.L. Bragg (1962), *The Crystalline State – A General Survey*, G. Bell & Sons, London, 1933, revised edition 1962

G.W. Brindley, K. Robinson (1945), 'The Structure of Kaolinite', *Nature*, 1945, vol.156, p.661

G.W. Brindley, K. Robinson (1946), 'The Structure of Kaolinite', *Mineralogical Magazine*, 1946, no.27, pp.242-253

G.W. Brindley, B.M. Oughton, K. Robinson (1950), 'Polymorphism of the Chlorites. I. Ordered Structures', *Acta Crystallographica*, November 1950, vol.3, pp.408-416 [clay minerals]

G.W. Brindley (1951), *Crystal Structure of Clay Minerals and their X-ray Identification*, Mineralogical Society, London, 1951

J.M. Broomhead (1948), 'The Structure of Pyrimidines and Purines. II. A Determination of the Structure of Adenine Hydrochloride by X-ray methods', *Acta Crystallographica*, March 1948, vol.1, pp.324-329

J.M. Broomhead (1951), 'The Structures of Pyrimidines and Purines. IV. The Crystal Structure of Guanine Hydrochloride and its Relation to that of Adenine Hydrochloride', *Acta Crystallographica*, March 1951, vol.4, pp.92-100

C.W. Bunn (1939), 'The crystal structure of long-chain normal paraffin hydrocarbons. The shape of the CH2 group', *Transactions of the Faraday Society*, 1939, vol.35, pp.482-491

C.W. Bunn, E.V. Garner (1947), 'The crystal structure of two polyamides ('nylons')', *Proceedings of the Royal Society, Series A*, 27 March 1947, vol.189, no.1016, pp.39-68

Dorothy Crowfoot (1938), 'The crystal structure of insulin. I. The investigation of air-dried insulin crystals', *Proceedings of the Royal Society, Series A*, 18 February 1938, Vol.164, no.919, pp.580-602

P.P. Ewald (1962), ed., *Fifty Years of X-Ray Diffraction*, International Union of Crystallography, 1962

M.S. Goldsztaub (1935), *Bulletin de la Société Française de Minéralogie*, 1935, vol.58, p.6 [lepidocrocite]

W.W. Jackson, J. West (1930), *Zeitschrift für Kristallographie*, 1930, no.76, p.211 [mica]

W.W. Jackson, J. West (1933), *Zeitschrift für Kristallographie*, 1933, no.85, p.160 [mica]

J.C. Kendrew (1948), 'Preliminary X-ray data for horse and whale myoglobins', *Acta Crystallographica*, December 1948, vol.1, p.336

J.C. Kendrew (1950), 'The Crystal Structure of Horse Met-myoglobin. I. General Features: The Arrangement of the Polypeptide Chains', *Proceedings of the Royal Society, Series A*, 7 March 1950, vol. 201, no.1064, pp.62-89

F.J. Llewellyn, E.G. Cox, T.H. Goodwin (1937), 'The Crystalline Structure of Sugars. Part IV. Pentaerythritol and the Hydroxyl Bond,' *Journal of the Chemical Society*, April 1937, pp.883-894

W.J. Lyons (1942), 'Structure of Cellulose as Revealed by Optical and X-Ray Methods', *The Scientific Monthly*, March 1942, vol.54, no.3, pp.238-246

H.D. Megaw (1934), *Zeitschrift für Kristallographie*, 1934, vol.87, p.185 [aluminium hydroxide and brucite]

H.D. Megaw (1935), *Zeitschrift für Kristallographie*, 1935, vol.90, p.283 [zinc hydroxide]

H.D. Megaw (1949), 'Hydroxyl groups in afwillite', *Acta Crystallographica*, December 1949, vol.2, pp.419-420

H.D. Megaw (July 1952), 'The Structure of Afwillite', *Acta Crystallographica*, July 1952, vol.5, pp.477-491

Helen D. Megaw (1952), 'Origin of Ferroelectricty in Barium Titanate and Other Perovskite-Type Crystals', *Acta Crystallographica*, 1952, no.5, pp.739-749

Helen D. Megaw (1957), *Ferroelectricity in Crystals*, Methuen, London, 1957

M.F. Perutz (1949), 'An X-ray study of horse methaemoglobin. II', *Proceedings of the Royal Society, Series A*, 3 February 1949, vol.195, no.1043, pp.474-499

J. Monteath Robertson (1948), *Journal of the Chemical Society*, 1948, p.1399 [resorcinol]

J. Monteath Robertson (July 1948), 'Bond-Length Variations in Aromatic Systems,' *Acta Crystallographica*, July 1948, vol.1, pp.101-109

W.H. Taylor, St. v. Náray-Szabó (1931), *Zeitschrift für Kristallographie*, 1931, no.77, p.146 [apophyllite]

A.V. Wendling (1938), 'An X-Ray Study of the Structure of Rubidium Dithionate', PhD thesis, McGill University, Montreal, Canada, 1938

W.H. Zachariasen (1934 - 1), *Zeitschrift für Kristallographie*, 1934, vol.88, p.150 [boric acid]

W.H. Zachariasen (1934 - 2), *Zeitschrift für Kristallographie*, 1934, vol.89, p.442 [oxalic acid dihydrate]

Bibliography B: Crystallographers – Selected Biographical Sources

U.W. Arndt, 'Charles William Bunn', *Biographical Memoirs of Fellows of the Royal Society*, November 1991, vol.37, pp.70-83

'George W. Brindley', *Journal of Applied Crystallography*, 1983, no.16, pp.217-8

Georgina Ferry, *Dorothy Hodgkin – A Life*, Granta Books, 1998

Georgina Ferry, *Max Perutz and the Secret of Life*, Chatto & Windus, 2007

Dorothy M.C. Hodgkin, 'Kathleen Lonsdale', *Biographical Memoirs of Fellows of the Royal Society*, November 1975, vol.21, pp.446-484

Oxford Dictionary of National Biography, Oxford University Press, 2004-2008: W.H. Brock, 'Sir Ernest Gordon Cox'; Soraya de Chadarevian, 'Sir John Cowdery Kendrew'; E.G. Cox, 'John Monteath Robertson'; Talab Debs, 'Sir William Henry Bragg'; Georgina Ferry, 'Dorothy Mary Crowfoot Hodgkin'; A.M. Glazer, 'Helen Dick Megaw'; Gill Hudson, 'Dame Kathleen Lonsdale'; Robert Olby, 'John Bernal Desmond'; Robert Olby, 'Max Ferdinand Perutz'; David Phillips, 'Sir William Lawrence Bragg'

'Dr. W.H. Taylor', *Journal of Applied Crystallography*, 1984, no.17, p.373

Bibliography C: Festival Pattern Group

C1. Archives

AAD – Archive of Art and Design, Victoria & Albert Museum: Helen Megaw Papers (AAD 1977/3/1-734)

DCA – Design Archives, University of Brighton: Design Council Archives - Festival Pattern Group Files (DCA 5384-1, 5384-2, 5384-3, 5384-4, 5395, 5396, 9466)

C2. Festival of Britain Guides

Science Guide: Dr. Jacob Bronowski, *Festival of Britain. 1951 Exhibition of Science, South Kensington. A Guide to the Story it tells,* HMSO, 1951

South Bank Catalogue: *Festival of Britain 1951. South Bank Exhibition. Catalogue of Exhibits.* HMSO, 1951

South Bank Guide: *Festival of Britain. South Bank Exhibition. A Guide to the Story it tells,* HMSO, 1951

Souvenir Book: Mark Hartland Thomas, *The Souvenir Book of Crystal Designs,* COID/ HMSO, 1951

C3. Contemporary Magazines

Architectural Review: Helen Megaw, 'The Investigation of Crystal Structure', *Architectural Review*, April 1951, pp.236-239

Architectural Review (December 1951): 'Crystal Structure', *Architectural Review,* December 1951, p.358

British Textiles: 'The Design Story of the Century', *British Textiles,* April 1951, pp.50-55

Design: Mark Hartland Thomas, 'Festival Pattern Group', *Design,* May-June 1951, pp.12-25

Homes and Gardens: Narcissa Wood, 'Designers Find New Themes', *Homes and Gardens*, May 1951, pp.34-35

Punch: Bernard Hollowood, 'Petri's Kaleidoscope', *Punch,* 8 August 1951, pp.167-168

Queen: 'Festival Pattern Group – Art and Science', *The Queen,* 25 April 1951

Skinner's Record: Margot Duff, 'Science Finds New Design Medium,' *Skinner's Silk & Rayon Record,* April 1951, pp.475-480

C4. Contemporary Newspapers

'Atomic and blood patterns are designs for carpets', *Banbury Advertiser*, 29 October 1951

'The "Atomic" Tie', *East Anglian Daily Times*, 16 April 1951

'Working in secret, local firm has produced unique Festival tie silk', *Suffolk Free Press*, 17 April 1951

'Crystal Patterns for Fabrics', *The Times*, 24 April 1951

C5. Books

Mary Banham and Bevis Hillier, eds., *A Tonic to the Nation – The Festival of Britain,* Thames & Hudson, 1976

Becky Conekin, *The Autobiography of a Nation: The 1951 Festival of Britain*, Manchester University Press, 2003

Soroya de Chadarevian, *Designs for Life – Molecular Biology after World War II,* Cambridge University Press, 2002

Elain Harwood and Alan Powers, eds., *Festival of Britain: Twentieth Century Architecture 5*, Twentieth Century Society, 2001

Lesley Jackson, *The New Look – Design in the Fifties*, Thames & Hudson, 1991

Lesley Jackson, *Contemporary Architecture and Interiors of the 1950s*, Phaidon, 1994

Lesley Jackson, *20th Century Pattern Design – Textile & Wallpaper Pioneers*, Mitchell Beazley, 2002

Angela Long and Shirley Wheeler, *Design4Science – The Visual Communication of Science*, Gmelin Press / University of Sunderland, 2007

Paul Rennie, *Festival of Britain – Design 1951*, Antique Collectors Club, 2007

J.A. Dunkin Wedd, *Pattern and Texture – Sources for Design*, Studio Publications, 1956

C6. Recent Articles

Soroya de Chadarevian, 'Reconstructing life. Molecular biology in postwar Britain', *Studies in History and Philosophy of Science, Part C,* September 2002, vol.33, no.3, pp.431-448

Sophie Forgan, 'Festivals of science and the two cultures: science, design and display in the Festival of Britain, 1951', *The British Journal for the History of Science,* 1998, no.31, pp.217-240

Lesley Jackson, 'X-Ray Visions', *Crafts*, September-October 2001, pp.32-35

Lesley Jackson, 'The Appliance of Science', *Crafts*, March-April 2008, pp.32-35

Tom McGill, 'Design Under the Microscope: The Festival Pattern Group 1951 - The Council of Industrial Design and the Mechanics of Industrial Liaison', *Decorative Arts Society Journal*, vol.31, 2007, pp.92 115

Mary Schoeser, 'The Appliance of Science', *Twentieth Century Architecture 5: Festival of Britain*, 2001, pp.118-126

Author's acknowledgements

Without the magnanimity of Richard Dennis and the amazing ingenuity and dedication of Sue Evans, this book would never have happened. I am extremely grateful to them both for taking this publication on board at such short notice and going out of their way to accommodate all my pernickety requirements. Thanks also to Alan Powers for designing the cover, and to Neil Dexter at Flaydemouse for pulling out all the stops to get it produced on time.

I am deeply indebted to James Peto and Emily Sargent at the Wellcome Collection for having faith in my proposal, and to the whole team at the Wellcome Trust for labouring so tirelessly to bring the project to fruition. Special thanks to Emily for her invaluable curatorial input and research, and to Nick Coombe and Graphic Thought Facility for their creativity in designing the exhibition.

I am grateful to all the museums, archives and other institutions and individuals who have arranged for material to be specially photographed for the book, particularly the V&A, without whose cooperation this project would have been impossible. I would like to thank all the exhibition lenders and everyone who has helped with my research and facilitated the publication, including: Joyce Baldwin; Professor Alan Mackay (Birkbeck College, University of London); Liz Broadley and Alex Findlay; Nick Coombe; Dr Lesley Whitworth and Dr Catherine Moriarty (Design Archives, University of Brighton); Wendy Lanchin (Design Council); Dr Simon Ford and John Davis (Design Council Slide Collection, Manchester Metropolitan University); Dr Maryon Dougill; Georgina Ferry; Dr Sophie Forgan; Clare Paterson (Glasgow University Archive Services); Michael Dacombe and Peter Strickland (International Union of Crystallography); Annette Faux (Medical Research Council Laboratory of Molecular Biology, Cambridge); Tom McGill; John McMeeking; Paul Johnson (The National Archives); Jeremy Farrell (Nottingham City Museums and Galleries); Vivien Perutz; Stephen Miller (Royal Festival Hall Archive); Susanna Rostas; Jennifer Kren (The Royal Society); Peter Morris, David Thompson and Tom Vine (Science Museum); Dr Judith Milledge and Dr Ian G. Wood (University College, London); Dr John E. Davies (University of Cambridge); Professor Lindsay Sawyer (University of Edinburgh); Professor A.M. Glazer (University of Oxford); Verity Andrews (University of Reading Library); Shirley Wheeler (University of Sunderland); Professor Robin Perutz (University of York); Andrew Henry (Vanners); Mark Eastment, Victoria Jarvis, Lesley Miller, Roxanne Peters, Sue Prichard, Suzanne Smith, Eric Turner, Eva White (Victoria and Albert Museum); Ann Wise (Warner Textile Archive, Braintree District Museum Trust Ltd.); Eric Whittaker; Katia Porter and Frances Pritchard (The Whitworth Art Gallery, The University of Manchester).

Special thanks to Ian Fishwick for all his help and support.

Picture Credits

W.L. Bragg, Atomic Structure of Minerals (1937): *fig. 27; Chapter 5: Source for Diagrams 8.10, 8.16, 8.28, 8.29, 8.30, 8.35, 8.53*
W.L. Bragg, The Crystalline State (1933, revised edition 1962): *Chapter 5: Source for Diagram 8.10*
Dr. John E. Davies (courtesy of Girton College, University of Cambridge): *FPG 50b, 68b-c*
Design Archives, University of Brighton: *figs. 4, 12, 43, 44, 47-48, 51-52, 58, 65; Diagram 8.15*
Design Council Slide Collection, Manchester Metropolitan University: *figs. 13, 18, 25-26, 29, 30-32, 38, 57, 60, 61, 63; FPG 45a, 77-80*
Exhibition of Science Guide (HMSO): *figs. 2, 11, 66*
International Union of Crystallography: *Chapter 5: Source for Diagrams 8.40, 8.41, 8.47, 8.48, 8.51, 8.55, 8.56, 8.57*
Professor Alan Mackay: *fig. 39, 49*
Medical Research Council Laboratory of Molecular Biology, Cambridge: *figs. 5, 15, 19*
The National Archives: *figs. 3, 14, 21-22, 33, 37, 40-42, 46, 50, 64*
Nottingham Castle Museum: *figs. 53-54*
Royal Festival Hall Archive: *figs. 10, 67*
The Royal Society: *figs. 6, 16, 20a, 24, 28; Chapter 5: Source for Diagrams 8.23, 8.25, 8.32, 8.33, 8.34, 8.39, 8.46c-d, 8.46e, 8.46f-g, 8.52, 8.54*
Science Museum: *FPG 10, 33-34, 45b, 46b, 47b, 48a, 50c, 73*
University of Sunderland (courtesy of University of Leeds): *FPG 53*
V&A Images / Victoria and Albert Museum: Cover (front and back); *figs. 7-9, 17, 20b, 23; FPG 1-5, 7-9, 11-14, 18-22, 24, 26-32, 35-38, 40, 43, 47, 48b, 49, 52, 54-59a, 60-63, 68a, 70-72, 74, 76; Diagrams 8.1, 8.2, 8.3, 8.4, 8.6, 8.7, 8.8, 8.9, 8.10, 8.17, 8.18, 8.23, 8.24, 8.25, 8.26, 8.27, 8.28, 8.29, 8.30, 8.32, 8.33, 8.34, 8.35, 8.38, 8.39, 8.45, 8.46c, 8.46g, 8.52, 8.53, 8.54c, 8.55, 8.56, 8.59a-b, 8.59c-d,*
Copyright Warner Textile Archive (Braintree Museum Trust Ltd.) All rights reserved: *fig. 28; Diagrams 8.14, 8.36, 8.37, 8.40, 8.43, 8.44, 8.46a, 8.47, 8.49*
J.A. Dunkin Wedd, Pattern & Texture (1956): *figs. 68- 69*
Wellcome Images (courtesy of Nick Coombe): *fig. 59; FPG 15, 17, 73*
Wellcome Images / Souvenir Book of Crystal Designs (COID/HMSO, courtesy of the Design Council): *figs.1, 34-36, 45, 55; Chapter 2; FPG 16-17, 23, 25, 41-42, 44, 50a, 51, 64-66, 69, 75; Diagrams 8.31, 8.48, 8.58, 8.59e*
Wellcome Images (courtesy of Professor Alan Mackay): *fig. 56; FPG 57b*
Wellcome Images (courtesy of Dr Judith Milledge): *Frontispiece, FPG 59b (p.60)*
Wellcome Images (courtesy of Vivien Perutz): *FPG 6*
The Whitworth Art Gallery, The University of Manchester: *fig. 62; FPG 46a*